COVENTRY LIBRARIES

Please return this book on or before
the last date stamped below.

PS130553 DISK 4

S

To renew this book take it to any of
the City Libraries before
the date due for return

Coventry City Council

keeping bees

a complete
practical guide

Paul Peacock

NOTE

This book is intended to give general information only. The publisher, author and distributor expressly disclaim all liability to any person arising directly or indirectly from the use of, or any errors or omissions in, the information in this book. The adoption and application of the information in this book is at the reader's discretion and is their sole responsibility.

An Hachette Livre UK Company
www.hachettelivre.co.uk

First published in Great Britain in 2008 by
Gaia Books, a division of Octopus Publishing Group Ltd
2–4 Heron Quays, London E14 4JP
www.octopusbooks.co.uk

ISBN: 978-1-85675-302-9

A CIP catalogue record for this book is available from the British Library

Printed and bound in Dubai

10 9 8 7 6 5 4 3 2 1

Contents

Introduction

If you've picked up this book, you may have been inspired by someone you know who has a hive or two, and you long to produce your own honey. Keeping bees is a most worthwhile pastime; you will make lasting friendships as you learn about bees.

Starting with bees

Of all human activities, the keeping of bees is one of the most satisfying, and it will change your life in ways you could not have imagined. It's a peaceful occupation, the ideal antidote to the fraught pace of modern life, and most beekeepers find that the hours they spend looking after their hives are deeply rewarding.

There are lots of reasons to keep bees, and increasing numbers of people around the world seem to be doing so. Some will be attracted to the products of beekeeping – the honey, wax and propolis, royal jelly and honeycomb – which are so expensive to buy and so useful in the home, medicine cupboard and kitchen.

Other people will be attracted to the social side of beekeeping and join a local group or branch of a national association. There is always something to be talked about – a problem to be debated, a success celebrated or a failure shared. Although it is possible to be a lone beekeeper, it is not advisable, and the large number of associations and clubs that are dedicated to the welfare of bees share an ethos based on the fact that every beekeeper was once nervous and new. In consequence, beekeepers are incredibly friendly and you are certain to receive a warm welcome.

Left: A WBC hive surrounded by flowering plants providing abundant nectar and pollen to feed the colony.

Perhaps one of the most surprising reasons for keeping bees is that they are likeable creatures. Few beekeepers ever fully overcome their childhood fears about stings, and it is sensible, of course, to be careful when you are around them, but the way that bees live together and organize themselves is endlessly fascinating, and soon you will find yourself comparing the human world with that of the honeybee.

The calming effect of beekeeping makes it an ideal antidote to the rush and bustle of the modern world, and at the end of the year you will be rewarded with anything from 20 to 100 jars of honey from just a couple of hives.

First steps

It is not possible to learn to keep bees with a book in one hand and a smoker in the other. By far the best way of learning is from someone else who has already had the experiences that you will soon be getting for the first time. The most important first step is to join a local group or association and to watch more experienced beekeepers as they work. You can then refer to a book to reinforce what you have seen.

This book provides a logical introduction to beekeeping, and it will give you the confidence you need to get your own hive or hives. Use it to reflect on what you have learned from other beekeepers and to remind you of what they do before you delve into your own bees. But more than anything, as far as beekeeping is concerned, read for knowledge but learn from your experience.

Are bees for you?

It is difficult to predict how you will react when you open a hive by yourself for the first time. However, there is much that you can do to prepare yourself for that moment, and you may be surprised at how easily a developing interest in bees becomes a lifetime's passion as you become intrigued by beekeeping.

Skills

Although you need no specific skills to keep bees, it is not something that should be undertaken lightly. It requires a serious attitude, attention to detail, a cool head and a willingness to learn.

You should not consider a lack of knowledge a bar to starting to keep bees, and you probably already have all the skills you will need. What cannot be learned, however, and what is perhaps the most important quality in a beekeeper is an ability to relax. An easy, calm manner will be your greatest asset.

The physical requirements – what you actually do in the hives, how you extract honey, feed the bees and check for the queen, and the dozens of other routine tasks – are easily picked up and are certainly no more complex than those required in the kitchen, the garden or your work place. They are all well within the capabilities of anyone who has the time and is ready to learn.

Time

Running two hives will probably take a couple of hours a week, although at key periods of the year, such as swarming time and when you collect the honey, you may spend a little longer. On the whole, however, bees prefer to be left alone. You should also be prepared, especially at first, to spend an evening each week at a local group or club where you will learn the ropes and meet other beekeepers.

Left: *Protected in an all-in-one bee suit, the beekeeper carefully removes frames for inspection, avoiding any sudden movements.*

Right: *A typical brood frame containing a lot of bees and at least two drones.*

Equipment

Most new beekeepers join a society not only to learn how to look after their bees but also to acquire the equipment and clothing they will need before they get their first colony.

The most important items are a bee suit, or at the very least a veil to protect your face, some beekeeping gauntlets or gardening gloves, a hive tool and a smoker. Secondary equipment ranges from some elastic bands, which you can use to keep clothes tucked in and bees out, a bee brush or a feather to remove bees from your bee suit, and a bucket to store the bits of waste from your hive inspections (see also pages 54–7).

Wherever possible it is better to buy new equipment, which will minimize the transmission of diseases, and if you join a national or local organization you will probably be able to get significant discounts on many items. Some of the more expensive and specialized equipment, such as honey extractors, are best shared and kept as a central resource.

As with any pastime, there are plenty of opportunities and incentives to accumulate a lot of equipment, some of which is not necessary, and over the years you will almost certainly pick up plenty of rather strange-looking odds and ends.

Space

A single beehive does not take up a lot of room – a space measuring 1 x 1 metres (3 x 3 feet) will be more than sufficient – and it is possible to have several hives quite close to each other. However, you will need somewhere to put your equipment, and it is sometimes said that for every hive with bees in it you will need one other as back-up. You will also need to keep spares of the elements that go to make up the hive – frames, chambers, supers and so on. A dry, secure garden shed is ideal.

Before you acquire a hive think carefully about its position (see also pages 60–3). Your bees need to be protected from people who might wish to harm them or their hive and should also be kept from accidentally harming your neighbours.

You may want to consider taking out some form of insurance, and your local bee society or association will be able to offer you the best advice.

The honeybee

Getting a worthwhile supply of honey from your hives depends on understanding both the life of individual bees and also how bees live and work together as a colony. It is worth spending a little time on learning about their lifecycle because a single hive will contain 30,000 or more individual bees.

The history of beekeeping

Honeybees live in well-ordered colonies made of individuals specially adapted for their role. There is a unity of purpose in the colony, and the whole mass of bees work to well-defined goals: from collecting nectar and pollen, to rearing new members to cope with the increasing workload as the season progresses.

Evolution

Honeybees have been around for over 120 million years, and they evolved alongside flowering plants, having found a niche by obtaining nourishment from pollen and honey, and in return pollinating the plants they visited. All the natural substances that plants use to prevent fungal and bacterial infections also work in the hive as bees incorporate the healing properties of the plants into the honey and wax they produce.

The hive, which is filled with high-energy food that would otherwise naturally spoil because of microbial action, is therefore able to remain practically disease free. And the healing properties of honey, pollen and propolis were so well known and recognized by the ancient Greeks that the scientific name, *Apis*, is derived from the ancient Greek word meaning healer. In ancient medicine, bee's ash (the material left behind when dead bees were burned) was said to have curative properties.

When the queen starts to run out of eggs the colony arranges for her replacement and makes sure that sexual reproduction takes place. The exchange of genetic material that takes place at this time is an important mechanism in the species' fight against disease, allowing any resistance that may have developed in one generation to be passed to the next.

Left: *A wild honeybee colony with folds of honeycomb.*

Early history

Prehistoric cave paintings that pre-date writing show humans taking honey from hives, and over 10,000 years ago both honey and wax were a welcome part of people's diet and lives. The discovery that honeybees retreat from smoke and gorge themselves on honey in preparation to evacuate the colony made a significant difference to early honey collectors, reducing the number of stings and making it easier to take the precious combs. Fire was a fundamental part of human development, and its application to enable people to remove honey from hives eventually became an important tool in bee farming.

The early hives were simple constructions – either a hollowed-out log with clay blocking the ends and with a small entrance on one side or a woven basket covered in clay – and simple log hives are still found in Africa today.

Beekeeping gradually became an important part of agricultural activities, and a land was prized for its ability to produce barley, emmer (wheat), cattle, milk and honey. It is little wonder that the production of valued foods became part of holy and religious rituals and writings. Meanwhile, the relationship between fire, smoke and spirituality came together in a number of ancient religions, including Hebrew, Mesopotamian and eventually Nubian and Egyptian.

The ancient Egyptians regarded beekeeping as part of the religious life of the nation, and priests used sacred smoke to control their hives. More than 6,000 years ago, the Egyptians believed that the sun would drop its tears as honeybees and that they would fall in the human world as honey and wax. The bee became the symbol of the country, and the pharaoh was named the king of bees and master of their spiritual essence.

Honey was expensive in ancient Egypt and was used only by the ruling classes. It was imported from all over North Africa and even from Asia. Honeybees, too, were traded, and the distribution of the local strains across Africa and Asia Minor and even into southern Europe (*Apis lamarckii*, *A. sahariensis* and *A. yementica*) today is a reminder of the ancient honey industry.

Above: *An ancient Greek example of a pottery skep.*

In China, where the first reference to honey was in the *Book of Chinese Medicine* written some 2,200 years ago, beekeeping mirrored the Egyptian methods. In India it has long been a Hindu custom to give a newborn child a taste of honey as a sign of the protection offered by the goddess Parvati.

Greek civilization was also influenced by the honeybee. It is little surprise that honey, called ambrosia, was regarded as the food of the gods on Mount Olympus. Ancient Greek hives – in the form of large, expandable, ceramic-lidded vases – have been excavated, and there are details of how they tended their bees in ancient literature. Hippocrates, the father of medicine, wrote about the culinary and medicinal uses of honey in the 4th century BC, and apiculture developed into a science that was studied as well as being part of secret religious rites.

The way that bees order their lives and the geometry of the honeycombs was incorporated into Greek thought.

The honeybee

In the 5th century BC the philosopher and mathematician Pythagoras believed that the regular patterns of the honeycomb were a reflection of something that pervaded the whole universe. It was also believed that bees' behaviour, the way they ordered their lives and the way they prepared themselves for winter had parallels in human life, and this was reflected in literature and in the development of democracy itself.

Greek methods of beekeeping – including the development of the lidded hive and the ability to keep colonies on a semi-permanent basis – were passed to the Romans, who subsequently took them around the known world as their empire spread. In the 1st century BC the Roman writer Virgil included a section on beekeeping in the *Georgics*, his writings on rural subjects. Virgil describes the siting and maintenance of an apiary, fighting swarms, the surrounding garden, the nature and qualities of bees, the gathering of honey, diseases of bees and the autogenesis of bees.

After the Roman period, beekeeping did not change much across Europe. Colonies were housed in straw skeps, which have their origins in the Greek pottery vases and are still sometimes used today, especially for catching swarms. Top-bar beehives, in which the bees draw honeycomb on sticks balanced across the top of the hive, are a development of the Greek style, and they have been common in Africa for 2,000 years and are still used by enthusiasts today all over Europe and the United States.

Recent history

The scientific study of bees, particularly the organization of the hive and the bees' industry, intensified in the 18th and 19th centuries. The relationship between the welfare of bees and the welfare of humans was increasingly recognized, and the vulnerability of the hive and the queen – especially through the winter – became topics of study and formed the basis of much allegory.

As communication between beekeepers in various parts of the world improved, discoveries in one country were gradually adopted in others. For example, sectioned hives were built in Russia in the 1820s, and soon after Russian beekeepers found ways of keeping the queen in a particular part of the hive.

One of the most important developments occurred in Philadelphia in the early 1850s, when the Rev. L.L. Langstroth wrote about the 'bee space', which revolutionized beekeeping. If bees are not happy with a space or gap in their living area, they will glue it up. Any space less than 4–5 mm (about ⅛ inch) across will be filled in with propolis (see page 33). If the space is a little larger the workers can get through, but the queen cannot. Rev. Langstroth also found that wax combs were never closer together than 7–8 mm (¼ inch), and he used these measurements to create a hive in which the queen could be confined to her own brood box while the other bees were allowed the freedom of the hive. Honey can, therefore, be saved in one part of the hive while brood can be developed in another.

The development of a pre-stamped honeycomb foundation in the mid-19th century was slow to catch on, but it eventually combined with the Langstroth hive (see page 44) to create almost exactly the modern beekeeping method. It became possible to harvest honey without brood mixed in with it and to take combs out of the hive without endangering the colony.

Since the modern hive evolved, there have been many changes in apiculture. It has become possible to manipulate a colony to suit the purpose of collecting large amounts of honey. In addition, it has become possible to move bees to take advantage of different crops, from fruits tree to setting honey from oilseed rape fields. A number of different types of hive have developed from the original Langstroth, and it is now possible to have a hive to suit different environments and styles of beekeeping (see pages 44–7).

Despite these innovations, the bees themselves have remained largely the same. They have reacted to threats from disease and adapted themselves to varying conditions all over the world. As a species they are equally at home in a skep, a top-bar hive, a modern hive, a hollowed-out tree trunk or even a dirty old chimney.

Bees have faced many threats in this period. Acarine (Isle of Wight disease) wiped out almost the whole of the British bee population in the early part of the 20th century, and this led to a great upsurge in bee research, which eventually resulted in the development of a strain of bee that was immune to the disease. The parasitic varroa mites (see pages 92–3) are but the latest threat to apiaries, and there are certain to be new ones in future.

Albert Einstein is reported to have said that if the bee disappeared off the surface of the globe man would only have four years of life left, and it is against this background that the lives of humans and bees remain inextricably intertwined.

The honeybee

Left: *A worker bee collecting much needed pollen in the springtime.*

Understanding bees

Bees have captivated peoples' imagination from the earliest times, and they will continue to do so. The closer we look into their lives the more fascinating they become.

Research and the honeybee

One of the problems facing any researcher into the life of honeybees is that it is all too easy to explain their behaviour in terms of human characteristics. Our preconceptions about them sometimes get in the way of our understanding.

The waggle dance

Some of the most interesting bee research was carried out by Austrian zoologist Karl von Frisch (1886–1982). A well-known entomologist, von Frisch's main work was a study of parasitic wasps, but he also worked on the life of the honeybee, and his ideas about the way bees communicate in the hive evolved into the idea of the 'waggle' dance, which is still debated by entomologists. Not everyone has been convinced by von Frisch's finding, largely because it sounds too human, and it is felt there might be another explanation.

Von Frisch discovered that bees communicate the location of sources of nectar by movements that are called dances, and the 'waggle' dance is a form of communication by which a bee can advertise to others in

THE WAGGLE DANCE EXPLAINED

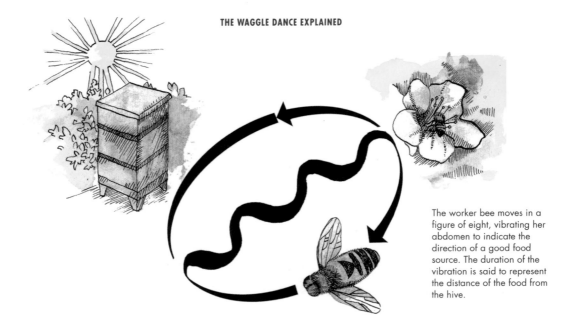

The worker bee moves in a figure of eight, vibrating her abdomen to indicate the direction of a good food source. The duration of the vibration is said to represent the distance of the food from the hive.

the hive the location of a promising food source. The bees will waggle from side to side for a short distance, the direction of their waggle indicating the bearing of the food in relation to the sun, and the duration of the waggle indicating its distance from the hive. Von Frisch found that bees are even able to take into account the different position of the sun as the day progresses. When he replaced the sun with a bulb the dance was disrupted and changed, but if even a small amount of blue sky was introduced to the experiment, the bees' dance was orientated correctly, which meant that they could see polarized light and use it for navigation purposes.

Von Frisch's researches also showed that bees cannot distinguish between some shapes but that they can distinguish between some colours, including ultraviolet (which is not visible to humans) but not red. Despite the on-going debate about the waggle dance, von Frisch's observations have become the standard methodology for research into bees.

Above: *Bees place pollen from the same plants together so that it is not mixed with other sources.*

Working together

More recent research into the life of colonies has found that bees sort food supplies according to colour and that they are able to differentiate between different food sources and decide, as a colony, which food is to be taken advantage of at any one time. It has also been found that bees can count, that they can elucidate shapes and that they have a wonderful ability to put right what they perceive to be incorrect in the hive. They will secure the defences of the hive and decide together which parts of the hive to work on first.

Overcoming problems

Much research into bee biology focuses on the problems faced by bees around the world. There are almost no feral colonies of honeybees in many parts of the world, and problems such as varroa and colony collapse syndrome remain serious challenges to colony numbers.

Traditional ways of dealing with varroa, for example, are no longer effective because the mites have developed an immunity to the treatments, and new strategies are constantly needed to combat them. Some recent research has been concentrated on the cell size. It is known that varroa mites prefer to inhabit the cells made by workers to house drones. These are slightly bigger than standard ones found in foundation comb, which can be either 4.9 mm or 5.1 mm (about ⅛ inch). It is not known if colonies that make all their own comb – top-bar hives have no foundation comb – have less varroa than more conventional hives.

Sadly, many bee populations around the world are under threat, and research has yet to find the answer. Human activity and interference may be part of the problem, however. Scientists have already interbred bees with questionable results, joining the aggressive traits of the African bee with South American bees to create a race colloquially known as 'killer bees'. Perhaps it is time for humans to start learning from bees rather than merely researching them.

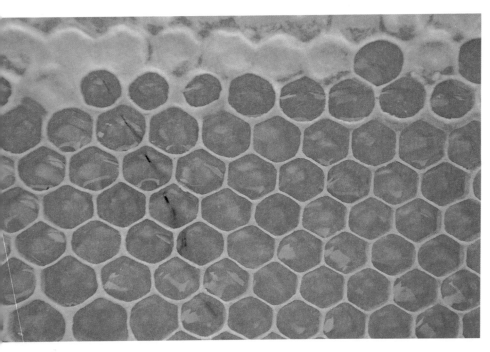

Bees and humans

Bees are also the subject of all kinds of modern research in many fields that are not related to their own activities but to their usefulness to humans.

Concorde

One of the most remarkable things about a beehive is the honeycomb structure of hexagonal cells. This has stimulated human imagination over the centuries and influenced research into topics ranging from the molecular structure of benzene to the production of new building materials. The floors of Concorde, the supersonic aircraft, were built like a hive's honeycomb but with a thin skin on either side, and this material proved to be extraordinarily light in relation to its strength. Honeycomb works in much the same way as a modern aircraft's wing. Much of its rigidity and strength comes from the contents being kept full.

Skin

The value of honey in the treatment of skin ulcers is also being researched. Widely used therapies have subjected the sufferer to months, if not years, of agonizing bandages, but the simple application of honey-impregnated bandages has shown promising results, reducing the healing time. The honey used is specially prepared and sterilized, and the boost to the local tissue creates new skin very quickly.

Hay fever

Research into the use of honey to treat hay fever is also giving promising results. The basic premise is that locally sourced honey is charged with pollen and pollen extracts from the patient's home area. This pollen, when ingested with honey, is thought to have the effect of mildly inoculating the immune system, relieving the symptoms of hay fever when pollen appears in spring and summer.

Arthritis

It is commonly said that old beekeepers do not suffer from arthritis, and research is now under way to find evidence about the effect of bee stings. Similar research into stinging nettles and the relief of arthritic pain seems to point in the direction of acetyl choline, a chemical found in bee stings and nettles.

Related research is ongoing into the few people who have a reaction when stung by a bee. This research is mainly linked to the immune system and how it responds to the chemicals in the sting.

Cellular chemistry

Honeybees are also being used for research at the cellular level, and scientists are studying the way honeybee cells respond to invasion, the bee immune system and the role of external bee products, such as propolis, on the internal chemistry of the honeybee. Researchers are learning a lot about the human immune system because of the way that bees fight disease themselves.

There are a number of pressures on bee colonies, particularly as governments begin to realize there is a growing decline in pollinators for agriculture. A lot of research focuses on the combined effects of modern life: pollution, insecticides, GM technology, global warming and on the abilities of bees to fight disease.

Because they have short lives and the colony is of paramount importance, bees put a lot of effort into fighting disease outside their bodies. The colony could be viewed as the organism, and the individual bees as simply a part of it. For instance, during the time nectar takes to ripen to make honey, there is plenty of opportunity for fungi to ferment this liquid. The worker bee mixes enzymes with the nectar to stop this fermentation from happening, the whole colony benefiting from the individual bee's immune system.

Left: *Honeybees are one of the main pollinators for plants and agriculture.*

The honeybee

Classification

The scientific or Latin name of the honeybee is *Apis mellifera* L., and like most botanical names it is largely descriptive. Like other botanical names of plants and animals, it is used around the world so that there can be no confusion from country to country about which particular species is meant.

Left: *Worker bees clean out the cell before the queen will lay a new egg in the bottom.*

Scientific names

The scientific name is simply a description of any living organism, so *Apis mellifera* means 'the honey-bearing bee'. The word *Apis* is the name of the genus to which honeybees belong, and it distinguishes them from other bees – bumblebees, for example, belong to the genus *Bombus*. *Apis* is derived from the Greek word for healer, and the Roman word for the bee was *apis*.

The species name, *mellifera*, means honey bearer, and it describes the honeybee perfectly. It derives from the Latin words *mellis*, meaning honey, and *ferre*, meaning to carry. Honey is called 'mel' or 'mell' in many languages – in French it is *miel* and in Italian *miele* – and we also find it in the word mellifluous.

The L that follows the species name stands for Linnaeus, the name of the person who described the species in the first place. Linnaeus, whose real name was Carl von Linné (1707–78), was a Swedish botanist and the originator of the modern binomial system of nomenclature of plants and animals.

Above: *Look carefully at a group of honeybees and you will see many individual differences.*

Scientific classification

The full scientific classification, or description, of the honeybee is:

CLASSIFICATION	NAME	DESCRIPTION
Kingdom	Animilia	which means they are animals
Phylum	Arthropoda	which means they have jointed legs
Class	Hexapoda or Insects	which means they have six legs
Order	Hymenoptera	which means they have membrane-like wings
Sub-order	Apocrita	which means they possess a sting
Super-family	Apoidea	which means they are bees
Family	Apidae	which describes them as social bees
Genus	Apis	Bee
Species	mellifera	Honey gatherer

Subspecies and hybrids

The species *Apis mellifera* (honeybee) is divided into about 30 subspecies and many more hybrids. Subspecies will mate with each other and produce viable, hybrid bees, which are generally unique in terms of colour, shape and behaviour.

Among the more important subspecies are *Apis mellifera carnica*, the Carniolan bee originally from Central Europe, which is docile and well suited to cool climates, and *A. mellifera ligustica*, the Italian bee, which is well mannered, rarely swarms and produces lots of honey. *A. mellifera mellifera*, the north European black bee, also known as the British black bee, was devastated by disease early in the 20th century. *A. mellifera scutellata* is the Africanized honeybee, which has a reputation for being more aggressive than the European honeybee, and *A. mellifera capensis* comes from the South African Cape. Within these races are many local sub-races.

The Buckfast bee is a hybrid of many bees from around the world, especially the European black bee and Italian bee, and was developed in Devon, in southern England, by a Benedictine monk, Brother Adam, in response to acarine, the disease that all but wiped out the British black bee.

Genome sequencing

The classification of honeybees around the world has been improved by the Honey Bee Genome Sequencing Consortium, a group of scientists who have worked out the sequence of genes in the DNA of various species.

The comparison of apine DNA has revealed much about the evolution of the honeybee and suggests that other insects have evolved more quickly than honeybees, which appear to have developed in Africa and spread around the world in two waves. Honeybees have fewer genes devoted to the immune system than other insects, a testament to the efficiency of their hive defence systems. The research has also shown that honeybees have over 160 receptors for smell, but only ten for taste.

Anatomy of the honeybee

Honeybees are among the most recognizable of all flying insects, and there can be few people who have not watched them buzzing from flower to flower in a summer garden. Most of the people who could identify a honeybee, however, probably know little about the insect's complex anatomy.

Body

Like all insects, a honeybee has a head, thorax and abdomen and six legs. Bees also have four wings, antennae and large, compound eyes. It is a remarkably successful arrangement, and there are more insects on the planet than any other animal.

The honeybee has no skeleton as such. Instead, it is held together by hard plates covered in wax, a little like a suit of armour. All the muscles are attached to the inside of the plates, and each plate is attached to the next by a tough but flexible membrane. The plates that make up the body are referred to as the exoskeleton, and without their waxy covering the insect would simply dry out and die.

Bees do not have blood vessels – they do not even have blood, for that matter. A fluid called haemolymph is pumped by a very simple 'heart' around the body cavity, transferring nutrients and oxygen to the tissues.

They do not breathe in the conventional sense and have no lungs. Holes in the exoskeleton, called spiracles, allow air into the centre of the body. When the bee has used up all the oxygen a large muscle rhythmically contracts and expands the abdomen to force air in and out of the spiracles. The spiracles lead to a system of tubes, called trachea, which carry gases to and from the centre of the insect. The trachea are bathed in haemolymph, and the rhythmic beating of the wings or the flexing of the exoskeleton maintains a constant flow of air in and out of the insect. A bee with blocked trachea cannot survive.

The thorax is largely filled with muscles and glands and with the nervous system, which stretches along the length of the lower part of the body. There is no brain as such, but a series of ganglia act as localized processors of inputs from sensors and send motor messages in response.

The gut is divided into many sections. The 'tongue' is a retractable proboscis, which mops up nectar and passes it to the honey stomach via a tube that runs through the thorax. Some partial processing of the nectar

Left: *The head, thorax, abdomen, wings and compound eyes can be clearly seen.*

takes place here, and the syrup is ejected at the honeycomb to make stores of honey. Beyond the honey stomach is the rest of the digestive system, a second stomach and the rectum for ejecting waste.

Head

Bees have two large compound eyes, one on each side of the head. Each eye consists of a few thousand simple eyes, each with a single lens. Some of the eye cells recognize colour and others movement. There are also three simple eyes, called ocelli, that recognize sudden shadows that fall over the bee. The first ganglion, sometimes referred to as a 'brain', processes the information from the eyes and creates the 'picture' the bee 'sees', which, remarkably, is not dissimilar to what humans see.

The antennae are packed with chemical receptors, which enable the bee to navigate around its environment. They are particularly important for recognizing the queen's hormones and for detecting any predators that enter the hive.

The mouthparts are adapted for drinking but do not bite. Honeybees also use their mouthparts to construct honeycomb, chewing and placing wax into position and laying down the antiseptic glue, propolis, which they use for structural security and hive health.

Legs

Honeybees have some of the most evolved legs in the entire insect world. The forelegs are similar to those of most other insects in that they are used for the important function of cleaning the antennae and eyes. In addition, the legs have sensory organs, mostly hairs, which enable bees to understand where they are in the world. Organs of balance, which are based on a system of hairs, are not dissimilar to the mammalian inner ear.

The hind legs of worker bees have a pollen basket or sac, the corbicula, in which pollen is taken back to the hive. The pollen is caught on the numerous hairs on the leg and transferred to the basket by the forelegs.

BEE ANATOMY

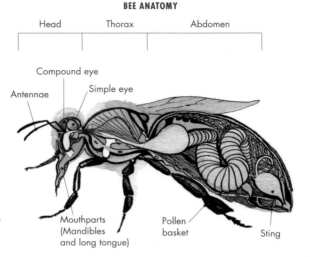

Head | Thorax | Abdomen

Compound eye
Antennae
Simple eye
Mouthparts (Mandibles and long tongue)
Pollen basket
Sting

Wings

Bees have four wings, and they differ from the diptera, mostly flies, whose second pair of wings are small or reduced in size, and from beetles, where one set of wings has hardened to create a case. The wings of a bee can beat independently in pairs, and it uses the flexibility of its exoskeleton to help flip the wing back into position, saving on the amount of work of the major muscles. Bees beat their wings billions of times.

The front and hind wing can be hooked together so they can act in unison if necessary. When the bee rests or is not using its wings the back ones fold neatly under the top ones, so it appears to only have two.

Sting

The bee's sting is made up of special reinforced plates with a venom sac attached. Once it has delivered its sting, the bee usually flies away, leaving behind the venom sac and a pair of pulsating muscles, which inject the toxin into the victim. The bee, meanwhile, simply dries out and dies shortly after because its body cavity is open to the air.

The honeybee

Types of honeybee

When you lift a frame from a hive and watch the bees, they seem to be a crowd of busy individuals. However, the whole colony makes up a super-individual, and each bee has a part to play in this 'collective life form'.

Queen

Because the average worker bee dies after about a month, she has to be replaced if the colony is to take advantage of the nectar and pollen the summer provides. The function of the queen, therefore, is to lay the eggs that will be tended by the workers and so build up the numbers of bees in the colony. A fully fertile queen can lay 2,000 eggs a day.

In addition to laying eggs, however, the queen has a number of other functions that are equally important to the colony. Without her special type of hormones, called pheromones, the workers would behave differently, becoming almost aimless and, above all, very angry. The queen's aroma, packed with pheromones, gives the hive its cohesion, and she maintains their loyalty – but only as long as she has eggs to lay. Once her fertility starts to fail the workers notice the change in her aroma and use some eggs to create replacements.

Many people think the queen bee is the ruler of the colony. This is not correct, for she is merely an egg-laying machine. In many ways she is inferior to the workers, and certainly her 'brain' is much smaller. She does not need the complex skills exhibited by worker bees and can usually be easily replaced by the hive. Should she die the workers can start to feed an egg with royal jelly and build up the cell into an emergency queen cell. A queenless colony will produce many replacements, and the first to emerge will often destroy the others. However, if winter comes and the colony remains queenless it will probably die out before spring.

Recognizing the queen

There may be a couple of thousand bees on each frame, and you will have to look carefully to find the queen. She is longer than the workers, a little lighter in colour and has a pronounced abdomen that looks full of eggs. Compared to a worker, the queen's body is tapered and elongated with smaller wings.

Making a queen

The queen is laid into a special cell known as a queen cell, of which there are different types (see pages 79 and 85). She is fed solely on a glandular secretion, royal jelly, which is packed with hormones and honey and makes the young queen grow quickly (see page 31).

Once the queen has emerged she will learn the layout of the hive and be tended by a group of workers, which are about two weeks old. Three days after she has emerged she will be ready to mate, and at this time she flies to around 10 metres (30 feet) into the air and is followed by a number of drones. She will mate with around ten males, whose genitals are pulled away, still pumping sperm into the queen, leaving the males to die.

She will then return to the hive to start her work of laying. Her sisters will have already prepared and cleaned a number of cells ready for her to use.

Worker bees

The workers, who make up the majority of bees in a hive, have different functions according to their age (see pages 28–9). Young bees that have just emerged from

the cell wander around the hive, cleaning and removing debris, while other workers police the brood frame.

Worker bees are not able to mate, but they retain their ovaries and produce eggs. If they are allowed to grow they will become drones, but only one in about a thousand drones is actually derived from a worker bee.

Recognizing worker bees

The worker is the smallest bee in the hive. She is shorter than the queen, has a less pronounced abdomen and is often more brightly coloured and strongly marked. Workers tend to have fewer hairs than the queen, but their anatomy is modified for carrying pollen and honey.

Drones

These male bees are squatter and squarer than both the queen and the worker bees, but are nearly equal in size to the queen. There are only ever a few hundred drones

Above: A marked queen laying eggs on brood frames with workers in attendance.

in the hive, and their function is to mate with the queen, which only a handful ever manage, after which they die.

Drones start life in a larger cell than worker bees and live for around three weeks. They do nothing around the hive, and at the end of the season they are thrown out, and no matter how much they hang around the hive entrance they do not get back in, eventually dying of hunger and cold.

Drones have no sting and tend to wander around the hive in a seemingly aimless manner, although they might be performing functions we know little about. You are most likely to see drones in the hive in late spring or early summer. Those that remain in autumn are expelled from the hive and left to die.

How the hive works

The colony's lifecycle is governed by the outside temperature and by the length of daylight. Bees are, therefore, more active in spring and summer, when they are seen visiting flowers and flying to and from the hive. Inside the hive the queen starts to lay more eggs and the overwintering workers begin to forage and feed the emerging new workers.

THE PARTS OF THE HIVE

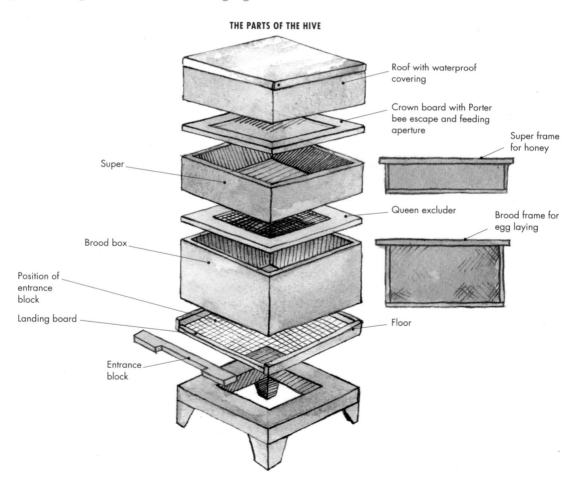

Roof with waterproof covering

Crown board with Porter bee escape and feeding aperture

Super frame for honey

Super

Queen excluder

Brood frame for egg laying

Brood box

Position of entrance block

Landing board

Floor

Entrance block

Inside the hive

The queen and the majority of the colony are found in the brood box, which is twice the depth of the honey boxes, also referred to as supers (see below and pages 48–51). She lays eggs in the lower centre of the brood frames and the workers deposit honey in a ring or arch on the outer, uppermost part of the frame. This is called the honey arch.

If you can see developing bees in uncapped (unsealed) cells you can be certain that there is a laying queen in the hive, even if you cannot find her. Once the grub pupates the workers cap the cell with wax. A similar process occurs in cells that are filled with honey: once the honey is ready it is capped to stop it from spoiling.

The queen excluder

A metal grid, wide enough to allow workers through but too narrow for the queen and drones to pass through, is used to keep the queen out of parts of the hive. The beekeeper can decide where to place the queen excluder, but in a new hive it is usual to allow the bees to fill all the brood frames first with brood and honey. The queen excluder is then used to separate this from a single super, where the workers are able to draw out (build) the honeycomb on the wax foundation sheet. Because the queen is excluded from these smaller frames they contain nothing but honey. Once this box is full the queen excluder is placed on the top of the first super, then followed by a second super.

This gives a hive with a brood box and a super of honey for the colony's winter supplies, while the queen excluder separates the hive from a second super, which will be for the beekeeper's use. In most circumstances the colony will need only the first honey-filled super.

Occasionally, when you are opening your hive, you will find that the queen has managed to get through the excluder and into the super, where she has laid eggs. Inspect the excluder carefully in case it is broken. The cause is usually a loose pin at the edge that enables the queen to push up the grill.

The brood box

The size of the colony increases as the season progresses, and all the bees will have to be accommodated at night because bees are one of the few non-mammalian animals to sleep. A full colony will form a 'bolus' or ball of bees in the centre of the brood box, where the hive is warmest, and the extras will spill over into the supers.

During the day house bees will take care of the grubs, they know the whereabouts of the pollen and honey stores and how to make the food they feed to the different castes of bee.

The crown board

Above the supers is a crown board and then a lid. The crown board is used to keep the colony intact under the lid. If you remove the lid of the hive it is difficult to handle the bees inside it – in fact, if the queen happens to be underneath it, it can be almost impossible. Moreover, the lid is difficult to remove if the bees have been able to propolize it (fill in the gaps with propolis; see page 33) because there is nowhere you can easily insert your hive tool to act as a lever. The crown board is easier to lever out if it has been propolized because there is access between this and the topmost super.

The landing board

Flying bees find it easier to gain access to the hive if there is a landing (or alighting) board fitted to the base. If there is no board, they fly to and fro at the entrance to the hive and have difficulty getting in, especially if the entrance has been narrowed, to keep the wind out or as an attempt to exclude robbing pests, such as mice.

Feeding

Any feeding should take place at the top of the hive, and it must be done so that the sugar syrup can be added without disturbing the bees too much (see pages 80–3). Although it's a fairly straightforward matter to remove the top of the hive and add some more syrup, do take care to wear your bee suit, even if you are just opening the lid.

The life of a worker bee

Most hives contain about 30,000 bees, the majority of whom are workers. The life of a worker bee covers three periods of 21 days each. It takes 21 days to grow from an egg to a newly emerged bee. Then the bee lives for 21 days in the hive, followed by a further 21 days as a foraging bee.

Left: *Eggs and larvae, known as grubs, in a healthy section of brood honeycomb.*

From egg to bee

Worker bees start life as a fertilized egg, which stands erect in the base of the honeycomb cell. As each egg develops over the next couple of days it falls over on to its side. The egg stage lasts for three days, and because the exoskeleton does not grow with the insect, there is a series of moults as the grub expands. After four moults, some ten days from being laid, the larva spins a cocoon and settles into its cell, which has now been capped. Eleven days later it emerges from the cell an adult.

At this point the now empty cell is cleaned and prepared by the workers for the queen to lay again.

In the hive

For the first couple of days the newly emerged worker finds her way around the hive, cleaning cells, removing debris and generally learning where things are. She locates the honey store and licks the queen, which secretes a chemical on to her that will inhibit the development of her sexual organs. A number of female workers will develop egg-laying capabilities, and these are usually recognized by the workers and destroyed.

Once she has learned where the brood and honey are stored, the young worker starts to feed the larvae, first the older larvae, where she can do least harm, then the younger ones. This period lasts for nine days. She will selectively feed workers, drones and queen larvae on the different regimes each caste requires. The worker bee's royal jelly glands are particularly well developed and peak in production from between six and twelve days after her emergence, and she will use the royal jelly to feed the queen.

In her final week as a house worker, the bee will build honeycomb, cap cells and deposit wax and propolis (bee glue). The cleanliness of the hive is the responsibility of bees at this stage of their development, and they must also maintain the structural integrity of the hive, where they tend to glue everything together with propolis. The worker's wax glands are particularly well developed at this stage. The instructions for creating the hexagonal cells in the comb is in the worker's genes.

Within a week of being a house bee the worker's venom sac will have filled, and she then switches to guarding the entrance of the hive, where she also assumes fanning duties to add to the ventilation of the colony. The airflow sends the queen's aroma into the atmosphere and acts as an attractant to foraging bees.

Outside the hive

At 22 days the worker bee embarks on a life of foraging for nectar, pollen and water. She will also act as a scout bee in case there is a swarm.

Her life as a forager begins with a few orientation flights, which enable her to fix the position of the hive exactly from different positions. There will be a number of these flights over the next few days, and if you move the hive as little as 2 metres (about 6 feet) from its original position, the returning bees will congregate where the colony used to be rather than finding their way back to the hive.

She will gather pollen from as far away as 3 kilometres (almost 2 miles) and travel many hundreds more in the next three weeks, and then, if she isn't lost, eaten, squashed, sprayed with insecticide or killed by a predator or disease, the worn-out worker bee dies after a busy life of less than two months.

The honeybee

Bee facts

- In her lifetime a worker bee will travel about 1,000 kilometres (620 miles)
- The brood box is maintained at a temperature of about 35°C (95°F), regardless of the outside temperature
- It takes 20 kilograms (44 lb) of honey to produce 1 kilogram (2 lb) of beeswax
- If bees are starving they will eat their own brood
- A bee's 'brain' is less than a cubic millimetre, but has a greater number of nerve cells per cubic millimetre than any other animal

Left: *Worker bees clean and feed the young grubs and eggs, then cap the cell when the grub pupates.*

Bee foods

It is the job of the older worker bees to feed the colony, and if you stand near the entrance to the hive you will see the comings and goings of the foraging bees. Natural bee food consists of pollen and nectar, collected from the nectaries of flowering plants. From these ingredients the bees manufacture all their requirements. They also drink water and a source should always be available.

Left: *A worker bee collecting pollen and nectar for the colony.*

Nectar

Nectar is a complex liquid made from a number of sugars manufactured by plants. These sugars are stored inside the foraging bee until she has enough to make it worthwhile to return to the hive.

The complex chemistry that turns nectar into honey begins in the honey stomach of the worker bees. This sugar is regurgitated into the mouths of the house bees, and the process is continued as they mix the solution for up to 30 minutes, adding enzymes and checking on the progress of the honey by taste. Once they feel that the honey is ready they evenly spread the liquid into the cells around the honeycomb so that the water in it can evaporate, thus helping to preserve it.

Nectar consists of about 80 per cent water, but honey is less than 22 per cent, a change that takes place in the honeycomb. Bees speed the evaporation of water by standing on the comb and fanning it with their wings. Once the honey is thick enough the bees put a cap on the cell, which remains in place until the stores are needed.

A colony may use over 50 kilograms (110 lb) of honey during a year, which is quite remarkable when you consider that another 20–30 kilograms (44–66 lb) is often taken by the beekeeper.

Pollen

The tiny specks of dust on the end of a plant's anthers are pollen – the male sex cells of flowering plants. When pollen touches the sap on the female flowering parts it grows until it fuses with an ovum to create a seed.

Pollen is packed full of protein and fat, and it makes an excellent food for young bees. It is usually mixed with water and the small amount of sugar in the bee's mouth, causing the pollen granules to 'grow' to make a substance that resembles bread. The processed pollen, which is often referred to as beebread, is stored in the honeycomb. Pollen is also used to add structural integrity to the outer parts of the honeycomb.

Bees collect pollen from a wide range of flowers, making at least a million trips in the course of a season. The pollen they collect is said to have beneficial effects for humans (see page 18).

Royal jelly

A rich secretion from the hypopharyngeal glands in the heads of young worker bees, royal jelly (among other substances) is used to feed the larvae in the colony, including workers. The amount of royal jelly fed to grubs depends on their caste: workers receive only a little during the first couple of days of life, drones receive more so that their sexual organs develop, but those grubs destined to become queens are fed on nothing else. Queens themselves are fed royal jelly as well as honey.

A simple analysis of the composition of royal jelly does not reveal its properties. It is mostly water (66 per cent), and protein and carbohydrate in equal quantities make up the bulk of the rest. A small number of minerals have been identified, but there are other substances that have not yet been recognized. It has recently been found that the composition of royal jelly differs throughout the year and that it might possibly have a role in reducing egg laying in autumn and the closure of the hive in preparation for winter.

Minerals and water

Bees need minerals and water in order to thrive. They extract minerals from the plants around them, but a water supply provided by the beekeeper is also important.

Artificial nectar and pollen

Bees frequently need feeding in order to get them through the coldest or leanest parts of the year (see pages 80–3). Feeders can be used to dispense house-hold sugar (sucrose) at a predetermined concentration, and the bees use enzymes to split this disaccharide molecule to create two simple sugars. However, bees that are fed entirely on sucrose can show symptoms of malnutrition. Bees are also sometimes given artificial pollen made from soya flour and other ingredients mixed with honey to make it palatable.

Most beekeepers become good botanists, and you will hear them talking of the lime flow, the apple flow or their equivalents around the world. In temperate climates there is a marked seasonality to the bee's year, depending on which plants are producing nectar and pollen. This is reflected in the fact that in the far north and south, the flows represent an increasing amount of plant activity whereas at the equator bee production is constant, mimicking the long season.

Below: *The Gresty waterer made with two upturned bottles in a trough filled with grit.*

Above: *Honey in the comb is prized by many over jar honey and can be eaten on toast.*

Right: *Local honey is said to be excellent for treating many ailments, including hay fever.*

Honey

This wonderful substance has been valued as the food of the gods from Egyptian times, and it has been a special part of mythology through all civilizations to modern times (see pages 12–15). Beyond honey's historical significance and wonderful flavour, it is magical food.

Honey is packed with carbohydrates, specifically equal proportions of glucose and fructose. These two sugars polarize light in opposite directions, and because bees can differentiate polarized light the honey is ready not only when it tastes right but when it looks right.

Honey also contains a small amount, usually about 2 per cent, of sucrose. The amount of sucrose permitted in honey is regulated and varies from country to country, although it is normally no more than 8 per cent.

Both fructose and dextrose are monosaccharides – that is, they are composed of a single ring of atoms – which makes them easily absorbable directly into the bloodstream and immediately available for metabolic use. Sucrose, on the other hand, is a disaccharide, and needs to be broken down in the gut before it can be used. Glucose is an ideal sugar to take because it can be taken by diabetics and is easily converted directly to glycogen, absorbed directly into the bloodstream.

Honey is acidic in nature, which has led to the belief that it contains some bee venom. This has been disputed by some scientists, but it has also been pointed out that the healing properties of honey cannot derive from the sugar component alone, and many people believe that venom must be present. The curative properties of bee venom have been widely researched, and they are said to help in a number of ailments (see page 19).

Another benefit of honey is that it is an excellent source of antioxidants, and research in the United States has shown that the darker the honey the better its curative qualities. The same research showed that honey was instrumental in slowing the production of LDL cholesterol, the so-called 'bad' cholesterol. Honey is beneficial as a part of a balanced diet, being a sweetener, a condiment and a preserver.

Honey is hygroscopic – that is, it takes water chemically from its surroundings – and for this reason it is frequently used as a preservative as well as a food in its own right.

Honey as bee food

Honey is used by the bees in the hive for all kinds of feeding. Grubs are fed honey, sometimes alone, sometimes mixed with pollen, sometimes with royal jelly. Sometimes the honey has to be diluted, and some water is also stored in the hive by bees for this purpose. Its main purpose, of course, is as a winter food, and the hive will need at least one super full of honey to get the bees safely through until spring.

Propolis

This substance is not actually a product of bees. Propolis is a waxy, resinous substance that is part of several plants' immune systems. Worker bees collect propolis from the plants as they forage, and then when they return to the hive they chew and manipulate it and use it to glue up their hives, filling any spaces that are less than 4–5 mm (about ⅛ inch) wide (see page 14).

Propolis is usually grey-brown to dark brown, and it can be found in any crack or scratch in the hive. When you pull frames out of the box you will find that they are often hard to move because they have been 'glued' in position by propolis, which eventually sets to become very hard. Propolis is quite brittle and cracks easily.

Because propolis originates in the leaf buds of a range of plants, from *Aesculus hippocastanum* (horse chestnut) to *Populus* (poplar) and various firs, it varies in content throughout the year as the plants from which it is gathered come in and out of season. Sometimes propolis is mixed with a little pollen, but it is always rich in aromatic compounds, from which it derives its useful antiseptic characteristics.

Within the hive propolis is used for its antibacterial properties, and it is known to reduce disease and parasitic activity in the hive. It is also used to improve

Above: *During the summer months bees work hard to collect all they need to get through the winter.*

soundproofing and reduce vibration, and any intruder that the bees are unable to drag out of the hive, such as a large insect, slugs and snails or even a mouse, will be completely covered with propolis.

It used to be said that bees used propolis to seal the hive against the elements, but modern research has shown that a colony will do well with increased ventilation in winter. Researchers are concluding that propolis is not used as a weatherproofing material.

Humans have used propolis extensively in holistic medicine for many years. It is said to be a remedy for sore throats, and some beekeepers believe that putting a piece of propolis in the mouth will help keep bacteria and viruses away. It is also said to be useful in the treatment of skin ulcers, inflammations, burns and scalds (see page 103).

Pollination

Honeybees are the major pollinators of crops all around the world. Their efforts increase agricultural yields by as much as 30 per cent and in areas where there is a shortage of honeybees, agriculture invariably suffers as a result.

Petal brightly coloured to attract insects

Anther bearing pollen

Stigma, where pollination takes place

Nectar (sugar syrup) is made here

Bees and plants

Bees are uniquely important as pollinators because they evolved in symbiosis with flowering plants. Although there are other pollinators – flies, moths, beetles and, of course, people themselves – and although some plants are wind pollinated, the majority of important crops are pollinated by bees.

When they are foraging bees actually work a crop, and the pollination is not a secondary consequence of pollen collection. The evidence for this can be seen in the higher percentage of pollination achieved by bees on crops compared with similar numbers of other insects. A

beehive on an allotment, for example, will increase the crop yield by more than 30 per cent, and this is seen no matter what crop the bees are working on. Traditionally, bees have been used to 'top-dress' fruit trees in spring, then have moved on to fields of oilseed rape in early summer, then on to flax, lavender, lupins, alfalfa and so on through the seasons.

Farmers pay a premium for the services of bees, and this fee is much more valuable than the amount of honey that is gained from the field. Commercial beekeepers combine these activities with rearing queens and producing the nucleus of new colonies. There have been several estimates of the value of pollination. In the USA it was estimated at $9.3 billion in 1987 and $14.6 billion in 2000. This takes into consideration the increase in crop yields should an alternative pollinator be necessary.

Around the world there are a number of areas where there are not enough pollinators available for crops, and the problem has become severe enough for the Food and Agricultural Organization of the United Nations to set up the International Pollinators Initiative (IPI). The problem is particularly severe in parts of China, Nepal and India, where the use of pesticides has made beekeeping increasingly difficult. Beekeepers have, naturally, tried to keep their bees away from areas where they know that crops are likely to treated with pesticides, but the result is that when the crops need pollinating there are no bees in that area to do the job.

How pollination works

Pollen is the male sex cell of flowering plants. It consists of fat and protein, but no sugar. As soon as pollen comes into contact with sugar it starts to develop a long tubule that has DNA at the far end. This DNA is designed to fuse with the DNA in the egg that is protected deep within the carpals of the fruiting body. Once this occurs the fruit will develop as it should.

Normally, pollen is produced well away from the female parts of the plant, and as bees crash around the flowers in search of nectar and pollen a high percentage of male sex cells from other plants of the same species will hit the sticky, sweet stigma, the foremost tip of the female parts of the flower.

Most flowers are arranged in whorls: first the sepals and petals, then the male sex organs (known as the androecium or male house). The inner whorl, which is often fused to make a single unit, is the female sex organ (called the gynaecium or female house).

It is impossible to guarantee which pollen goes to what plant, but it is statistically evident that enough pollen of the right type falls on to the plant's stigma for successful pollination to occur.

Right: *A honeybee searching for nectar and pollen.*

Avoiding stings

Humans have been hard-wired to panic when they see an orange and black insect coming towards them. Indeed, this tendency for alarm has been so embedded into millions of years of evolution that some insects with no stinging mechanism replicate bee markings in the hope that we and other animals will leave them alone.

Trust your kit

Bee suits are designed to stop bees from stinging, particularly if you wear a good layer underneath (see page 54). The net visor will keep bees out of your eyes and away from your face, and if you are afraid – quite reasonably – of bees stinging your head and neck through the cotton suit, wear a hat.

The only possible exposed part might be your hands, but you can even cover these with garden gloves or beekeeper's gauntlets. Even though you might have seen some beekeepers dealing with their bees with bare hands, it doesn't mean that you will be a better or poorer beekeeper if you follow suit. Don't be unnecessarily macho, no one likes being stung.

Once you are covered you can begin to relax and to enjoy the relationship between you and your bees.

Using a smoker

The most important piece of equipment you have is your smoker, which, used correctly, will cause the bees to recoil and fill up on honey. Their extended stomachs make it less likely that they will sting. Using a smoker is described on pages 73–4, but a small puff at the entrance will warn the bees that you will be entering the

Left: *A smoker with its fuel: old pine cones, dried rotted wood, and grass.*

Right: *Working in the hive is comfortable when you trust your kit and have a routine.*

hive. Remove the lid and give another gentle puff. The aim is to waft the smoke over the bees in a single motion.

Smells

Bees do not like unusual smells and will attack anything that has a strong smell. Beer or garlic on your breath, perfume and cologne, sweat, hairspray – all of these can make bees skittish, and even though they will not all fly towards you in a frenzy, you will notice more of them flying into your visor and there will be more buzzing around your head.

Jewellery

Bees often fly straight towards anything that shines, and if you are going to work with bare hands you might want to remove any rings you're wearing and your watch. Bumps and folds in clothing, often caused by watches or bracelets, can become places where bees congregate, and if they get trapped they can attempt to sting.

The hive

Do not stand in front of the hive entrance, where the bees fly in and out, because when you have finished you will find that your back is covered with foraging bees that have returned and landed on you. It is not unusual to see a couple of hundred bees on the back of an unwary beekeeper's suit.

Don't panic

If you are a newcomer to beekeeping it is a good idea to enter a hive for the first time with an experienced beekeeper, which will at least give you the experience to realize that there is no need for panic around a large number of bees.

Even if you do get stung, which is likely at some point, keep calm. It's best to remove the sting with the flat of your hive tool. Don't squeeze the bee because you will only force more venom into the wound. The venom is packed with histamines, and the sting will feel a little like a mild burn. Although it is an unpleasant sensation, it is not unbearable.

Some people say they will allow the stinging bee to twist itself off the skin, stopping the sting from becoming dislodged and killing the bee. This excessive attention to the bee is not necessary: get the bee and its sting off you as calmly and as quickly as you can.

Some people react badly to stings. An antihistamine tablet often helps, but if you feel short of breath or dizzy,

Left: *If you stand in the way of bees at the entrance, expect to be covered within a short time.*

Above: *A bee sting still pumping venom is best removed with the edge of a hive tool.*

Dealing with a sting

If you are stung, do not squeeze the bee, which will only force more venom into you. Slide it away with a finger nail or your hive tool. Take an antihistamine tablet if necessary and if you get any symptoms, such as shortness of breath or dizziness, seek immediate medical advice. Some people recommend putting honey on stings, but for most people the pain will last for maybe five minutes and then pass.

especially after the sting has stopped hurting, seek immediate medical advice.

Why do bees sting?

Drones have no stings, and although queens have stings, they use them only to kill other queens, and they have hardly ever been found to sting anything else.

Worker bees do sting. The sting has a barb on the end, and so each bee can sting only once unless you are prepared to allow her to writhe on your skin to set herself free. A worker bee only ever uses her sting to protect the colony, never herself, because delivering the sting usually ends in her having half her abdomen ripped away as she is swatted away.

Bees sting mostly when they are at or near the hive, and it is actually a reflex action, which the bee performs when it feels threatened. The idea that what is bad for the individual bee is bad for the colony controls the bee's reaction. If you kill a bee at the hive it will release a pheromone that will make the other bees more likely to sting. The same pheromone is produced by bees that see you waving your arms around, that smell your perfume or sweat, that hear you making loud noises around the hive or that see you wearing dark clothes. You will notice the bees fanning pheromones to alert more and more

bees, and if you shake down bees from a frame a sudden burst of activity will take place.

Bees will warn you before they start stinging. They will fly at your face and whizz around your head. A high-pitched vibrating in your ears is a warning that you are considered to be an invader.

'Killer bees'

In parts of North and South America Africanized honeybees are regarded as a problem because what seems like the whole colony will attack. They are similar to the European honeybee in that each bee will sting only once, and they are not possessed of powerful venom. They will warn you in the same way as European honeybees, and if you are not wearing a bee suit the best thing to do is run away. The bees will follow you for up to 200 metres (about 220 yards), but you should, just about, be able to run more quickly than they can fly.

Cover your face if you can and do not go into water, because the bees will simply wait around for you to come up for air.

It is unlikely these bees will survive in Europe if they are ever introduced because they cannot survive cold winters, but climate change makes their future spread uncertain as temperate region winters become warmer.

The hive

In the wild bees build nests in inaccessible places – the highest branches of trees or on overhanging rocks – but the beekeeper's aim is to provide his honeybees with somewhere they can live, visit plants and produce honey, so that the beekeeper can then conveniently remove the honey without disturbing the bees. To this end, humans have developed various styles of hive.

The hive

The hive is not just a place where bees are kept, it is a way of controlling, feeding and harvesting and good beekeeping involves becoming proficient in using it.

What do you need from a hive?

When you are purchasing your first hive there are a number of factors to take into consideration. Importantly bees come first. If they are not happy or healthy in the hive then there is no point in keeping it. Secondly, the hive must be easy to work with. This means it has to be light enough to lift around yet strong enough to take the weight of all that honey – which is much heavier than water. A super will weigh around 12–15 kilograms (26–34 lb) and you need to be able to lift this without damaging yourself.

Most hives are anchored with a restraining strap these days, unless they are in a very sheltered position, which enables hives to be made from lighter materials. PVC hives have excellent insulation properties and you get more brood for your money with them, but without strapping they are likely to blow away in a high wind.

You should consider how many pieces the hive breaks down into. When you are piling up supers inside outer covers, and having to deal with the bees at the same time it is easy to get confused about how to reassemble it.

Standardization

Beekeeping has been made easier by having equipment and accessories of the same size and specification. For this reason it is best to use similar equipment to other beekeepers in your district or club. This way you will be

Right: *A single compartment large skep hive, in use today as it has been for many hundreds of years.*

Above: A row of modern hives in an apiary. Notice how the heights indicate the size of each colony: some with many honey-filled supers, others with one.

Right: A well-kept WBC hive in a productive herb and flower garden. The attractive style of the WBC hive looks ideal in a cottage garden.

able to get a ready supply of boxes and feeders, queen excluders, floors, lids, foundation, frames and crown boards. In the UK the National hive is most common. In America and many parts of the world the Langstroth or the Commercial hives are most popular. However, in most countries a large number of hive types are available. See pages 44–7 for further information.

Types of hive

There are several types and sizes of hive, and your choice will largely depend on what is available and what is used in your area (see pages 64–5). They range from simple skeps, which can be used to capture swarms, to more complicated arrangements with several supers.

The main problems with many types of hive are that it is not possible to inspect individual sets of brood, it is difficult to locate the queen and the honeycomb is mostly a mix of brood and honey. The ability to take out individual frames of honeycomb is one of the features of the modern hive, and it was developed not, as many believe by the Rev. Langstroth, but by a Russian beekeeper, Petro Prokopovych, who also invented a simple wooden queen excluder, made of wood and with holes drilled in it. Modern hives are all designed around the minimum bee space, 4–5 mm (about ⅛ inch), which is the room needed for a worker to move comfortably around the hive. Bees will use propolis to glue up gaps smaller than the bee space.

Skep

This traditional hive is basically a large straw or wicker hat-shaped container with a hole in it for the bees to get in and out. A skep has no base, and the bees hang from the 'roof', from which they draw (build) their honeycomb. The bottom of the skep is completely open, but the bees do not seem to be troubled by the extra ventilation in the winter. Even if you have a skep you are unlikely to use it in the long term, although it can be useful in an emergency for capturing a swarm. There are many designs of skep, depending on the requirements or location. Skeps are remarkably warm in winter and are exceptionally good for harvesting propolis.

Top-bar hive

The bees' ability to draw their own honeycomb from a stick or any overhang has long been used to the advantage of beekeepers. Top-bar hives provide nothing more than a bar from which bees draw down their own honeycomb. They are, in fact, only a little more complex than a straw skep, and you get honey as it used to be before modern hives were developed. There is no physical way of separating the brood from the honey stores, but by moving bars around it is possible to keep the brood nest at one end of the hive and the stores at the other.

You cannot spin the honey out of the honeycomb using an extractor. Instead, it has to be allowed to drip into a bucket or is pressed and filtered. Many people eat the honeycomb and honey together.

Langstroth hive

The Rev. Langstroth collated information on the distance between frames, the bee space (see page 14) and optimum hive dimensions, and he came up with a style of hive that is now used almost everywhere in the USA. It has a deep brood box that contains ten frames and shallower supers.

Right: *Two beehives with double brood boxes from a commercial apiary.*

Right: *The WBC hive can still be bought but are not used as much in recent years because of the cumbersome nature of the outer boards.*

National hive

The National hive is the one most often used in Britain and Europe. It is similar to the Langstroth in that it consists of separate boxes for the brood and honey. The opening of the hive allows the frames to be stacked either parallel to the opening or at 90 degrees, which alters the path of air: parallel frames are known as the 'warm way'; angled frames are known as the 'cold way'.

The frames are flush with the top surface of the box, and the bottom surface is deeper than the end of the frames, creating a bottom bee space. The bees are able to crawl over the top bars of the frames because of the space afforded by the frames above.

WBC hive

The William Braughton Carr (WBC) hive is what most people think of when they imagine a hive. It has pagoda-type frames that create an outer wall for insulation. The WBC takes the same frames as the National hive, but it fits only ten frames. It is less common than other hives, partly because it is more expensive and partly because it is inconvenient to have to dismantle the outer covers. When moving the WBC hive, the outer boards take up a lot more space than a National or Commercial hive. A number of variations of the WBC hive exist that are especially designed to help bees cope with consistently poor weather conditions or strong winds.

Dadant hive

This type of hive is fairly similar to the Langstroth, and you sometimes hear beekeepers talk about Dadant–Langstroth hives. It has ten frames and is largely similar to the Langstroth in that almost all the parts are interchangeable. The largest of the hives, the Dadant is found in France and some parts of Spain, but is not used in many other parts of the world.

Smith hive

This hive holds 13 frames that are of the same dimensions as those used in the National hive. It is a top bee space hive, which means that the frames come flush with the base of the box (see page 46), and there is a gap at the top for the bees to move around the frames. It is little used by newcomers to beekeeping today, but the extra frame makes it a popular style of hive with those who do still have one.

Commercial hive

This type of hive has the same dimensions as the National. It is available as either top bee space or bottom bee space and is often used with brood boxes above the queen excluder. The size of this hive is used to create a strong colony, which will produce a large volume of honey in the supers. The supers of Commercial hives are frequently just brood boxes on top of the queen excluder, allowing the colony to produce up to 23–27 kilograms (50–60 lb) of honey per box. Commercial beekeeping is a heavy job during harvest! It is not uncommon to see a Commercial brood box with National supers because of their lighter weight.

Right: *The lid of a National hive being removed to show the crown board beneath, with notes attached to the board.*

Parts of a hive

The modern hive has evolved over many years and will continue to change in response to the challenges that new pests and the requirements that a developing honey industry brings. What is now considered best practice is very different to the rules followed a decade ago, and the same will apply, after another decade.

Above: *The mesh of a varroa floor ensures that mites that fall from the bees actually exit the hive completely.*

Stand

This has many functions in addition to raising the brood box off the floor. By the time your honey supers are full they will be quite heavy, and making sure that the top brood box is at a suitable height will make lifting them as convenient and comfortable as possible.

Floor

Many beekeepers still use solid wooden floors of the correct dimensions for their particular hives. However, research into the ecology of the hive has shown that varroa mites that fall on a solid floor can climb back into the hive and reinfect your bees. An open mesh floor allows the mites to fall out of the hive, and bees seem to thrive with the increased ventilation.

The floor also provides an alighting point for bees to land on when they are returning to the hive.

Brood box

The brood box is the chamber where the mass of bees congregate around the queen and where she lays her eggs. She will lay where she feels the temperature is right, and this tends to produce a large semicircle in the bottom half of the brood frames. There are usually 10, 11, 12 or 13 frames in the brood box (depending on your type of hive), and the brood nest is usually centred on the middle frames, but then spreads out to fill the whole chamber.

Queen excluder

This is an ancient invention (see page 27). Pieces of wood with holes the size of a worker bee drilled in them have been used in the earliest of the modern hives to keep the queen in a particular section of the hive. Because the queen is fatter than the workers, she is unable to squeeze through the holes.

Modern queen excluders usually take the form of a metal grill, and they fit on the top of the brood box.

Top bee space or bottom bee space?

Your hive will be either a top bee space (with a space above the frames) or a bottom bee space type (with a space below the frames), and you will notice that there is a space underneath the grill of the excluder. You must remember to position the grill so that bee space is maintained for your particular style of hive.

The bee space varies in size according to the bees themselves. In order to understand the bee space you should visualize what is going on in the hive. Bees on adjacent frames need to be able to work back to back in order to get inside cells.

Clearly the drone and queen are much bigger than the worker and this is reflected in the design of the queen excluder, which has a space that is large enough for the worker, but too tight for the queen. The gap on the excluder is 4.3 mm (about ⅛ inch), but a damaged one might have a 'hole' in it – a bent bar allowing a space enough for the queen to squeeze through.

The space also has a bearing on the way bees protect the hive. During the winter, bees will often narrow the gap around the nest site to as little as 7 mm (¼ inch) so individual bees can crawl through but predators find it more difficult to enter the hive.

Of course, in the case of the top-bar hive and skep, the concept of the bee space means nothing to the beekeeper – but the bees themselves will create comb that conforms roughly to bee space principles for security and weather protection.

Petroleum jelly

Bees tend to use propolis to glue up everything within the hive, and separating surfaces can be awkward. If you cover abutting surfaces with petroleum jelly or a solution of liquid paraffin and beeswax, the bees will not glue these surfaces, making them easy to separate and work with.

Above: *A new brood box with clean frames, all ready for the introduction of a new colony.*

Above: A smart super filled with new honey frames early in the season for the young colony to make wax cells for honey storage.

Supers

The supers are usually half the depth of the brood box and consequently have only half-sized frames in them. They are normally reserved for honey stores, and a queen excluder is usually placed between the brood box and the super so that no eggs will be laid in the honey frames, assuming the queen excluder is not damaged.

A mature colony in the height of the season might have several supers of honey – not all of which are destined for the colony.

Crown board

This is a board that covers the whole of the top of the highest super, and it is simply a device to keep the bees at bay when you have removed the lid of the hive. There are usually a couple of holes cut into the board through which you can feed the bees by means of a feeder, or you can place a Porter bee escape into the hole. This is a device that acts like a one-way valve, allowing bees out of the hive, but not in again, and forcing them to re-enter through the bottom.

If you have no queen excluder you must check under the crown board for the presence of the queen when you lift it from the hive.

Some modern crown boards do not have holes in them, and they are simply a covering. (It is an excellent place for keeping your notes on the colony.) You can get crown boards with a bee space on either side; otherwise, you will have to position the board to preserve the bee space.

Lid

The lid of the hive, which must be waterproof, usually has a metal covering. Some lids have ventilation and escape holes, others do not. The lid overhangs the top super, and these two together form a locked unit. Some beekeepers paint identifying shapes on the lid.

Strap

If your stand is fixed to the ground secure a strap around all the boxes and over the lid of the hive, passing it through the stand, so that it will keep the entire hive secure in gusting winds and also from vandals, who have been known to turn over hives. A strap is really important if you are using a polystyrene hive, which is particularly light and susceptible to strong winds. The nylon strap with metal locking device has revolutionized beekeeping. It is completely secure and strong enough to anchor the whole hive to the stand – assuming the stand is in turn anchored to the ground. One drawback is that the cheaper ones can have clutches that corrode, making it a bit of a problem opening a hive after a long absence after the winter.

Above: *Fixing a hive with a strap. Notice the box below the floor and brood box. This is simply to raise the hive to a comfortable height.*

Caring for a wooden hive

Wooden hives should be scorched before use with a blowtorch to kill any pests or diseases that lurk in the grain and corners, and the wood should be intact, without any cracks or warps.

As with most aspects of beekeeping there is some debate about painting hives: some beekeepers say bare wood is best, because this allows condensation inside the hive to evaporate naturally. Others say that the paint causes the wood to swell and eventually dislodges nails and joints. Many beekeepers, however, paint their hives inside and out, or at least treat the wood with preservative. It is a good idea to paint the hive in colours that do not stand out too much, particularly if it is in an urban area. Exterior-quality paint is adequate, as is wood preserver, but you must make sure that the hive box is very well ventilated and thoroughly dried before you introduce the colony.

Frames

The hive

Movable frame hives, a little like modern top-bar hives, have been around since at least the late 16th century. They made it possible to remove individual combs for inspection. Better still, most of the bees remained undisturbed. Modern frames are designed to be easily and quickly removed.

Modern frames

Modern frames are quite complex in design yet simple in construction, and you will need to buy frames that will fit into your hive. British National frames fit National, Smith and WBC hives, or you can buy Langstroth, Dadant and Commercial frames.

Frames are available in two sizes – one for brood boxes and the other for supers – and they hold a template for workers to build the comb. The template, called foundation, is stamped beeswax – bees will not make up on any other substance, such as plastic. Foundation was invented in 1857 by Johannes Mehring,

and it took several years to take off but has now become standard. It is mostly bought as worker-sized cells, but you can also get the larger, drone-sized foundation.

The foundation has a wire running through it, and there are two types. In one the wire forms an M, and in the other the wires are laid singly, appearing at the top of the sheet. In the M-style foundation, which is common in Britain, the sheet goes into the frame so that wires create an upright M. The single-wired sheets are more usual in the USA.

Like everything else in the modern hive, the frames are designed to preserve the bee space.

Assembly

A frame has several parts and it has to be assembled before it can be used in the hive. Although the size and style of frames vary from hive to hive, the basic principles remain the same for each. A wired, pre-pressed wax sheet with a template for the cells is sandwiched between pieces of wood that maintain the bee space within the hive and allow the frame to be easily removed for inspection.

Left: *A super frame with pins holding it together at the most strategic points.*

How to assemble a frame

A frame has several parts and it has to be assembled before it can be used in the hive. Although the size and style of frames vary from hive to hive, the basic principles remain the same for each.

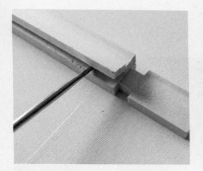

1 Use a penknife or similar to prise out the loosely attached strip of wood in the top bar (it will eventually be nailed back into position). Take care that you do not split the wood.

2 Hammer a couple of 19 mm (¾ inch) gimp pins to hold the bars in position. The top bars and sidebars will fit snugly together with the cut-out slits to the inside.

3 Assemble one pair of the bottom bars to create a recess in which the foundation will sit.

4 Take the wire that comes out of the bottom of the sheet of foundation and cross it back over the surface of the sheet.

5 Pin back the trapping bar into position and assemble the other bottom bar to make a perfect frame.

6 Pin all the joints and make sure the whole assembly is firmly joined together.

The beekeeper's toolkit

Beekeepers need a wide range of tools, equipment and special clothing to protect themselves from bee stings. You can purchase this equipment through your local beekeepers' club.

Bee suit

An all-over bee suit is one of the most important items you will need. Not only will the suit stop bees congregating in the folds of clothing or up your trouser legs, but it will add an extra layer of fabric so that bee stings are not able to penetrate to your skin if you have reasonably thick clothing under it. The built-in veil will stop bees attacking your face and vanishing down your ears, which is a thoroughly unpleasant experience.

You should launder your bee suit as often as possible so that infections are not passed from hive to hive, or even from apiary to apiary.

If you wear gauntlets make sure that the sleeves of the bee suit are on the outside to avoid any creases in which bees can congregate. Gauntlets do not afford complete protection, and in any case you really do need to be able to feel with your fingers as you inspect the hive. For this reason many beekeepers prefer to wear latex gloves, changing them regularly to maintain good hygiene around their hives.

Boots

It is a good idea to wear sturdy boots that will give you a good purchase on the ground. It is vital that you do not slip or fall over when you have a frame of bees in your hands. Make sure that your bee suit forms a seal over the boots so that bees cannot congregate around the tops of your boots and find their way into your socks.

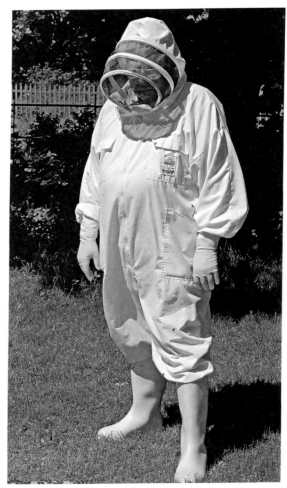

Right: A well-clad beekeeper in bee suit and veil, Wellington boots and gloves.

Above: *A well-stoked smoker producing lots of cool smoke.*

Smoker

A good-quality smoker is likely to be the most expensive piece of equipment you will buy. You can burn a range of materials in the smoker, from wood shavings and paper to dried pine cones and grass clippings, but make sure that you take care of the bellows and do not allow them to get too wet, when they sometimes rot.

There are several styles of smoker, but they all operate on the same principles: you fill the chamber with paper and a material, such as old wood, and once it is properly alight you close the lid. Avoid over-vigorous pumping with the bellows or you will end up with flames coming from the funnel. You are aiming to deliver a cloud of non-acrid, cool smoke.

Above: *Smoker fuel can be anything from old wood to pine cones or grass.*

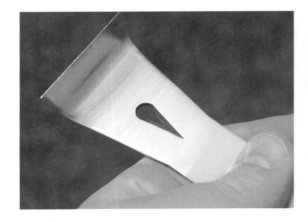

Above: *The hive tool with a sharp hooked edge. The hole is excellent for removing bee stings.*

Above: *The hive tool is a wonderfully designed piece of equipment. Here, it is levering a heavily glued-in frame.*

Hive tool

This small tool is a design masterpiece. There are two types, one with a J-shaped end and the other a scraper with a curved end. Some beekeepers like to have both types, but one or the other is usually sufficient, at least at first. You can use the tool to scrape away comb and queen cells from the side and bottom of frames, to separate and lift frames that have been glued together with propolis, to clean hives, to inspect cells, to hold bees and to destroy queen cells. As you gain experience with your hive you will continue to find new uses for it.

Bee brush

A bee brush is mostly used for removing bees from your bee suit. It's easiest to do this if you have someone with you to check the back of your bee suit. A few feathers will do just as well, but the brush has long, soft bristles, which allow you to move the bees without harming them or making them angry.

Note pad and pen

You will find it enormously useful to take notes about what you have done in the hive and when, and if you

Above: *A bee brush is ideal for moving bees. Here the beekeeper is coaxing a swarm into a cardboard box.*

have more than one hive it is absolutely vital. Record when you applied varroa treatments, and in some countries you are obliged by law to make a note of the manufacturer's batch number. Remember to be vigilant with your record keeping.

Useful items

The items described on pages 54–6 are essential, and you cannot even think about acquiring a swarm until you have bought them. There are, however, several other useful, but not essential, pieces of equipment, which you can obtain more gradually.

Disinfectant

Use a solution of a proprietary non-aromatic disinfectant to clean your hive tool, especially if you go from hive to hive or from apiary to apiary. If you are a member of a beekeepers' association there are umpteen opportunities for passing infection among members' hives and your personal stock.

Drawing pins

Keep a small stock of drawing pins to mark frames when you are working in the hive so that you do not forget where you last worked.

Forceps

Surgical grasp forceps are extremely useful and can be used to hold something shut so that you can handle it without having to keep up the pressure. They are particularly useful when you are dealing with queen cells because you can simply grab the cell, squash it and remove it in a single action.

Cap

Even though you will wear an all-over suit that includes a veil, it sometimes happens, especially if you are bending over, that the veil is pressed against your face, leaving you vulnerable to being stung through the mesh. In order to hold the veil away from your skin, it can be helpful to wear a peaked cap of some kind.

Bee paint

You will need to mark the queen with brightly coloured paint so that you can easily recognize her on the frame. This will considerably reduce the time you spend handling the bees (see also pages 88–9).

Rubber or cloth squares

The squares are cut so that they are the same size as the open top of the hive. You simply place a square over the exposed frames while you are working on something else or have to move away from the hive. The square keeps the bees calm and protects them from the sun or a sudden shower.

Bucket

You will find that a bucket is one of the most useful things you can have. Take it with you when you visit the hive and use it to hold the trimmings of wax that you might render down to make items from candles to soap.

Scissors

Some beekeepers trim the queen's wings so that she is unable to fly far from the hive, and if you do this you will not lose all your bees in the event of a swarm.

Queen cage

This is not something that a new beekeeper is likely to need, but you might be given a queen in a plastic mesh cage, plugged at the open end with some sugar candy or even newspaper. Introducing a new queen into a hive can be difficult – the workers might turn against her – but when she is slotted between the frames in the brood chamber while she is still inside the cage, the workers have time to lick the queen without harming her and are more likely to accept her. After a few days, when the workers have recognized her as their queen, you can either release her yourself or leave the workers to chew through the candy or newspaper to set her free.

Getting started

The most important question a new beekeeper must consider is where to site his hive so that the bees are safe from attack from both humans and animals and can, in turn, do no harm to neighbours or passers-by. You will probably have an ideal spot in your garden, but assess it carefully from everyone's point of view, and remember that once you have introduced the bees into the hive you will not easily be able to move it (see pages 90–1). Only then can decisions about acquiring a hive and finding a reliable source of a new colony of bees be made.

Where to keep bees

It is possible to keep bees almost anywhere – in ordinary gardens, on rooftops or balconies, beside agricultural crops, at sea level or up a mountain. Bees will happily live in the most unusual places, and in the wild they are more likely to be found in the cavity of a wall or in an empty chimney or unused drainpipe than in the newest, most sophisticated hive that money can buy.

Above: *A WBC hive on a garden patio with plenty of woodland cover behind.*

Choosing a site

The ideal site for a hive is, of course, a garden that is next to a field or an orchard, but as gardens have become smaller and houses closer together this is possible for fewer and fewer of us. Most beekeepers have to make the best of the conditions in their own garden, but with a little thought there is no reason you cannot keep a hive or two in even a comparatively small space.

When you are looking at the opportunities for positioning a hive or two in your garden, think first of all about the protection that your existing fences and hedges will afford the hive. Sheltering the bees from prevailing winter winds should be a priority, and a dense evergreen hedge is ideal. A high hedge or fence will also encourage your bees to fly higher, thus avoiding the heads of passers-by or neighbours. Although some beekeepers recommend positioning hives so that the entrances face south (in the northern hemisphere), in a sheltered garden this is less important than making sure that the openings face away from the prevailing wind.

Avoid low ground in the garden because in winter cold air tends to sink and gather in 'frost pockets', where the air temperature can be several degrees lower than in the rest of the garden. This will chill the hive and could lead to the death of the colony.

If possible, avoid siting your hive directly under a large tree. Not only might strong winds cause branches

and twigs to fall on the hive in winter, when your aim should be to keep your bees as quiet and undisturbed as you can, but also rainwater may drip on to the hive. However, in the height of summer a little shade over the hive will be welcome, so check the orientation of your garden and position the hive near to a deciduous tree or large shrub so that there is some shelter from the overhead midsummer sun but also so that any winter sun will fall on the hive and warm it.

Finally, make sure that you have easy access to the hive and that you have plenty of room to move around when you are working on it. Remember that you will need space around you to place the boxes that you remove when you inspect the hive, and you also need somewhere safe to put your smoker so that smoke doesn't constantly drift into the open hive.

Urban beekeeping

More people in the world live in urban areas, so the majority of us will be urban beekeepers. There are special problems associated with keeping bees in this type of environment, and the chief of these is pollution, which can degrade or even kill colonies. Sites near a factory or a main road are never going to be suitable, and if the only garden you have is affected by pollution but you would like to keep bees you will have to think about renting a site away from your house or keeping a hive on an allotment if you are lucky enough to have one.

Surprisingly, perhaps, the tops of tall buildings are good places for bees, although you will have to take extra care to protect the hives from strong winds and to make sure that you can take effective precautions against swarming: the cost of removing a swarm from someone's heating system could be prohibitive.

Unfortunately, humans can also be a problem. Your neighbours might simply be afraid of bees, so screening the hive with a high hedge that encourages your bees to

Urban bees

An important aspect of urban beekeeping is to be sure that you have gentle bees. Ask your supplier about the temperament of your colony. The gentlest bees, those who return to the hive quickly after they have been disturbed, are the best for bees that are kept in an urban garden.

fly high (see page 60) might be all you need do. Avoid looking at your bees when there are people about – especially when children are playing near by or when neighbours are sunbathing in their garden. When you do visit your hive make sure that you subdue the bees with your smoker and cover cloths and do not spend too long looking at them. Try not to shake dozens of bees out of the hive and keep them inside as much as possible. Take the time to explain to your neighbours how the hive works and how they can avoid being stung, and, of course, don't forget to offer them a jar or two of honey to thank them for their cooperation.

Vandals are, sadly, a fact of life. Try to choose a position for the hive that is out of sight of casual passers-by, and always secure the hive with a strap.

If your hives are at ground level make sure they are completely stable. Local pets, particularly cats, are quite heavy enough for a misjudged leap on to a hive to knock the whole colony on its side.

Providing nectar and pollen

Your bees will need a good supply of nectar and plenty of pollen from a wide range of plants to provide an adequate supply of food from early spring to early autumn. The availability of food is referred to as 'flow' – bees collect nectar and pollen as it flows from plants – and there are three main periods during the year. (See also pages 116–21.)

Below: *Oilseed rape is a favourite summertime crop for beekeepers as it is a good source of pollen.*

Spring flow

Although it is known as the spring flow, in most areas this period actually begins in late winter and lasts until early to mid-spring.

This is a time when fruit trees, including apples and cherries, are beginning to bear blossom, and some orchards still have their fruit 'top-dressed' (pollinated) by visiting bees. In addition, your bees will visit trees like *Salix* spp. (willow) and weeds, such as *Taraxacum officinale* (dandelion).

In later spring, if you live near or on a farm *Brassica napus* (oilseed rape or colza) and early legumes, such as beans and peas, will be important food sources. Oilseed rape is especially attractive to bees, but the honey produced crystallizes quickly, and remember that farmers are, of course, likely to protect their crops by spraying them with pesticides, which will harm your honeybees.

In built-up areas, although there will be a few early-flowering shrubs, such as viburnum, and spring-flowering bulbs, such as *Galanthus* (snowdrop), you will probably have to supplement your bees' food with syrup (see pages 80–3).

Summer or main flow

Leguminous plants, from clover to peas and beans, are a good source of pollen in the country, and in more urban gardens trees such as *Tilia* spp. (lime) produce pollen-rich flowers.

A great opportunity for beekeepers is the flow on the heather moors that lasts from midsummer until well into autumn. To take advantage of this, however, you must be able to move your hive, so avoid the WBC style.

Autumn flow

In late summer to autumn many garden plants bloom in sufficient quantities to provide enough food for your bees. As well as different types of heather (ling) and all the many garden plants, *Hedera* spp. (ivy) are an invaluable food source.

Getting started

Pesticides

Like all insects, bees are susceptible to being poisoned by pesticides, so you should not keep your hives where there is any likelihood that they will be exposed. Bees will, of course, travel a couple of miles when they are foraging for food, so it will not be possible to keep your bees away from all the possible dangers, but you should do your best to avoid droplet contamination.

Below: *During spring, flowering fruit trees are an excellent source of pollen.*

Acquiring a hive

Beehives are available from many sources, and your main difficulty will be in choosing the style you want and deciding whether to buy new or second-hand. Before you buy anything, however, talk to other beekeepers to hear about the pros and cons of the different styles and, if possible, arrange to see some of them in situ so that you can see how other beekeepers position and manage their hives.

Housing your bees

The main problem in keeping a hive of bees is that, to avoid losing your bees to a swarm, you need to have a couple of hives available, and the appropriate number of supers, as well as the associated frames, bases and stands. There may, therefore, be far more expense involved than you had at first thought.

Perhaps the best time to start a first colony is in the period after swarming, which is usually from mid- to late summer. This will give your new colony time to build up over the rest of the summer without the imminent threat of swarming. You will then be able to acquire the necessary kit over the winter months so that you are ready for the following season.

To house a single colony of bees in the new season you will need:

- 2 brood boxes
- 22 brood frames
- 2 queen excluders
- 2 floors and hive lids
- at least 2 super boxes with 22 super frames

Left: *A new National hive on a galvanized stand set in concrete, with a brood box, super and closed crown board, but waiting for its lid.*

Of course, you can get away with a single hive, which requires half this number of items, as long as there is someone from whom you might obtain a ready supply of extras if you need them.

A standard set-up for a growing colony at the height of the season will include: a brood box with associated varroa floor on a stand, a queen excluder and at least one (preferably three) supers with the appropriate number of frames. The first super will be full of honey for the colony's own stores to see them through the winter. Once it is full you can move the queen excluder above it and wait for the next super to fill, and so on.

What to buy

It is probably best to use the style of hives that are most widely available and in use in your area (see pages 44–7). If you are learning about beekeeping at a club or association it is advisable to buy something that you are already familiar with.

Second-hand equipment

It is only worth buying second-hand equipment if you know the conditions in which it has been stored and are certain that everything is in good condition. You must be sure that the hive has been thoroughly cleaned and

Above: A group of beekeepers learn the ropes under the supervision of experienced tutors – the best way to learn about beekeeping.

scorched with a blowtorch before you put your new colony anywhere near it.

To avoid the transmission of diseases it is best not to buy second-hand frames. These are more difficult to clean and are awkward to dismantle to make sure that even the joints are completely disease and pest free.

The main incentive for buying second-hand equipment is cost. Modern polythene beehives are much cheaper than their wooden counterparts, and their use is increasing, especially among commercial beekeepers. However, they must be strapped to a stand that is secured to the ground – these hives are so light that you cannot just put a brick on the roof to stop it blowing over.

Many beekeeping associations sell equipment, and buying from them not only makes sure that you get a good deal but profits will go towards the club, encouraging beekeeping in your area. Better still, you will get a lot of impartial advice for free. There is also a worldwide network of beekeeping suppliers, most of whom are interested in the welfare of the bees themselves, so ask questions and get as much information as you can with your purchase.

Setting up a hive

Before you even think about acquiring a colony of bees, you must set up the hive correctly. It is almost impossible to move the hive once the bees are in residence, so take the time in the early stages to site and set up the hive so that you can plan carefully for the introduction of the bees.

Positioning the stand

Position the hive so that the entrance faces away from strong prevailing winds so that rain will not be blown into the interior. Make sure, too, that there is space for you to stand at either side of the hive while you work, rather than directly in front of it, so that bees can return from foraging without landing on your back.

The stand is an important part of the hive. It raises the working area and keeps the base off the ground. Hives that rest at ground level quickly fill with all sorts of pests, including slugs and snails.

Ideally, choose a stand that is made from tubular steel, which can be cemented into position. This allows you to hold the whole hive together with a strap, which is especially useful if you are using one of the modern, lightweight, polystyrene hives, but will also prevent the hive from being knocked over.

Cold way or warm way?

To allow the air to circulate freely, set your frames the 'cold way', at an angle of 90 degrees to the entrance. To reduce air circulation, set your frames the 'warm way', parallel to the entrance. Use the warm way if your hive is in an exposed site and the cold way if it is in a sheltered position.

Left: A mesh floor is now used as part of a varroa mite reduction strategy.

Establishing the nucleus

Bees that are bought usually come in a nucleus, often known as a 'nuc'. This is a box with four or more frames of brood, some stores of honey, a marked queen and a host of attendant workers. Your first task is to place the frames of brood in the centre of the brood box and pack in new frames on either side.

When the bees are in the brood box, place a super on the top and a feeder with 2:1 syrup solution (see pages 80–3). Then you should leave the hive alone.

After a week you can take a peek inside to see if the queen is laying and if there is evidence of worker bees flying about and collecting honey and pollen.

As the grubs appear in the brood you will notice that the new frames are being drawn out built with comb and that the size of the colony is increasing. You can then place a queen excluder and a super of honey frames on top of the brood box. This super, when it is full of honey, will represent the bees' winter supplies.

You can place another super on top of this – this will be your honey – only when the bees have filled the first super with honey. You must make sure that you cater for their needs before you think about your own wishes.

Below: *A super with ripening honey and bees drawing cells on the crown board. This colony might need a new super for more room.*

Acquiring a colony of bees

If at all possible, buy your bees from a reputable breeder. Every association will be able to put you in touch with several suitable breeders, and only by acquiring bees in this way will you be able to plan for the introduction of the bees. Decide where you are going to keep them, where you will put and organize the hive and so on before you even think about buying the bees.

Choose your bee type

The best way of making sure that you get the most suitable bees is to talk to other beekeepers – ideally several. You need to choose a hybrid or type of bee that will do well in your own particular environment. You might, for example, have plans to take your bees to various locations to take advantage of the flow of honey in different areas, and you will, therefore, need a bee that is suited to the range of conditions it will encounter.

As a new beekeeper you will need a colony that will put up with your mistakes, will not surprise you with untoward swarms and mess up the hive with lots of brace comb (wax bridges between adjacent surfaces). Above all you are looking for well-behaved, gentle bees.

Before you buy, ask the supplier:

• Are they good mannered?
• Do they swarm easily?
• Is the queen prolific?
• What race or subspecies of bees do they come from?

You will be able to assess their suitability for your apiary and conditions by asking your supplier who else has this type of bee. You might also want to know how many colonies the supplier collects queens from, so that you can be confident with their techniques, and remember to

ask if there have been any recent problems, information that will minimize the likelihood of your bees having any diseases.

It is possible to order a nucleus through the post. Make sure that you are around and able to respond the moment the bees are delivered.

Timing

Make sure that you have everything ready for your bees before you buy them. You do not want to leave them packed into a brood box for very long, and 24 hours in hot temperatures might kill them.

Consider whether you will need help in putting your bees into the hive – if it is the first time, having an experienced beekeeper on hand will give you the confidence you need even if you don't ask for their assistance – but you will need to coordinate the dates.

Also, make sure you have a good supply of food for the bees. It might take many days for bees to collect nectar and pollen, and even longer to make this food available for the colony, so the golden rule is *feed them*.

Accepting a swarm

Occasionally you have no option about setting up your hive. Someone descends on you with a box of bees because they have heard that you wanted some and they were given a swarm – and here it is!

If you are a newcomer to beekeeping get help when it comes to taking on a swarm. You might be taking a weak swarm, a swarm of an old queen that might not last the winter, or a set-off of diseased bees. New beekeepers should only accept a swarm if they have reliable back-up. Nevertheless, housing a swarm of bees can often prove successful – but keep your expectations on hold a little until the colony is well established.

Swarms are usually taken by placing a box or an old-fashioned skep over the swarm or, if the bees are on a

Above: *A nucleus box with four brood frames ready to place in a new hive.*

branch, by cutting off the branch or shaking the swarm into the box or skep (see pages 86–7).

The simplest way of introducing a swarm into a hive is to shake it from the box into a brood box half-filled with brood frames. However, it is more interesting to 'throw' the swarm into the hive as described on page 87. Bees always walk upwards into the hive.

Bee care

As with all animals that humans have 'domesticated', from pet mice to horses and ponies, the beekeeper is responsible for his bees' wellbeing and for making sure that they have sufficient food and drink and that the hive is secure and well made. The beekeeper is also responsible for keeping his bees healthy and free from pests and diseases, and the best way to do this is to establish a system of regular checking. Once you have become used to your bees, however, you will find that this routine becomes an enjoyable experience.

Inspecting the colony

Part of your summer routine will be to inspect your bees to assess their health and to make sure that the queen is laying eggs and the worker bees are making honey and capping cells. You should check the hive at least once a week in spring and summer, and then reduce this to once a fortnight as the days begin to shorten and temperatures drop in early autumn.

Above: *Regular inspections of the hive in the summer are important, but they should be of short duration.*

Checking the hive

You can, of course, tell if a colony of bees is healthy simply by checking inside the hive, but you should be able to get some first impressions before you remove the lid. You should see a large number of bees – a handful every few seconds – flying in and out of the hive. They should have a determined path and alight on the board and march straight inside. You will not be able to see if the bees are full of honey, but the pollen sacs on their back legs should be visibly full of yellow, orange or white pollen, and they will look dusty.

If you knock on the hive and get the response of loud buzzing and the bees appear haphazard and fairly angry it is likely that the queen is dead or missing.

Opening the hive

You should only enter the hive when the temperature is above 15°C (59°F), it is not raining heavily and the wind is light. Do not chill the bees at all.

Avoid opening the hive more frequently than is absolutely necessary. Bees are not pets, and if you examine them too often you will lose them. In summer, when the bees are active, you should check them once a week, but in late spring and early autumn once a fortnight will be sufficient. In winter leave your bees alone, although you should, of course, continue to check the outside of the hive for security and signs of predation.

Looking inside the hive will normally give you an indication of the colony's strength, how much food they have or might need and the health of the queen. You will be able to judge if you need to take any remedial action, add a super, give some food, remove queen cells, re-queen and so on. You will also be able to check the colony for signs of disease.

The first time you open your own hive can be a nerve-racking moment. Many newcomers to beekeeping like to ask a more experienced person to accompany them – just in case. Keep calm and prepare yourself in a logical way. Put on your bee suit, find your smoker and make sure you have matches or a lighter. Put the spare frames and supers close to hand.

When you light the smoker avoid pumping the bellows as if it were a steam train. It will get too hot, and the fuel will burn too quickly. Give a couple of short puffs away from the hive to make sure that it is still alight, and only then should you approach the hive.

As you open the hive take care that you do not accidentally kill any bees. When a worker bee is in distress she gives out an alarm pheromone that will be ignored by most of the bees. However, if several bees are injured or distressed the alarm increases, which will have the effect of making the other bees more prone to attack. You will notice bees flying straight at your visor. Try to stay calm if this happens, and try not to breathe heavily because your breath will excite them even more.

The usual set-up of a hive is to have a super on top of a queen excluder and then a super for the colony's stores and then the brood box. You might have added a queen excluder above the colony stores super if it is full of honey, and the queen is not likely to lay in this.

How to open and check the hive

Opening the hive should become a ritual that you always stick to. A good routine saves mistakes and trouble later.

1 Make sure your smoker is properly lit and delivering good quantities of smoke. Remember not to stand directly in front of the bees' entrance, then puff the bellows to deliver a little puff of smoke into the hive from below. Do this before you remove the lid. The smoke warns the bees of fire, and they will start to take honey, which makes them more rotund and less likely to sting (though it is by no means certain).

2 Remove the lid and put it upside down to one side of the hive. You will stack the other components on top of this, except, perhaps, the crown board and queen excluder, which you might like to place on the other side of the hive.

3 Deliver another puff of smoke underneath the crown board as you lift it. If there is no queen excluder in the hive, carefully lift up the crown board and look on the underside to check that the queen is not there. If you forget to check and quickly put down the crown board there is a danger that you will kill the queen (this advice applies equally to the lid of the hive and to the queen excluder itself). If you see the queen carefully coax her back into the brood box.

4 Because the queen occasionally gets through the excluder (see page 27) and lays in the smaller honey frames check these frames before you remove any of the top super boxes to make sure there are no eggs or grubs in the cells.

5 Use your hive tool to lever a super from the hive. Do not try to lift it by the frames. Instead, grasp the sides of the box firmly and lift it up, keeping your back straight.

6 Gently prise up one edge of the queen excluder with your hive tool. Never try to force it or bang on the side of the hive to loosen it. Check the underside of the excluder for the queen. If there are a lot of bees on it, shake them off into the hive with a deliberate single sharp movement. Repeat this if they do not come off. Lay the excluder down by the side of the hive.

7 Carefully remove the brood frames and check for the queen, healthy grubs and capped brood cells.

Left: *Making notes is important – the best place to keep them is inside the hive itself, especially if someone else is to enter the hive at a later date.*

In the brood box

When you have opened up the brood box look for the queen. A gentle puff of smoke will send the bees on the surface of the frames scurrying out of the way so that you can gently lever out the frames. Hold each frame over the brood box as you inspect it so that if the queen falls off she will land back in the box.

If you can find the queen so much the better. This is much easier if she is marked, which has the additional advantage that you can colour code the marking to indicate her age (see pages 88–9).

If you can see lots of eggs in the cells then you can be sure that the queen is there – unless, of course, there are lots of drone cells, which might mean the queen is running out of sperm. In general, however, a large number of eggs in the height of the season is a sign of a healthy colony.

If the colony is strong you will be able to see many bees, all of them busy and moving with purpose. The laying queen will be surrounded by workers, and you will find that each of the frames has a central brood area

with capped and uncapped cells. Surrounding these are the storage cells with capped and uncapped honey. There will also be plenty of cells containing pollen, which is used for food. These cells are uncapped, and usually only one type of pollen is stored in each cell – you will be able to see the orange, yellow and white accumulations, which are often the work of a single bee.

If you have one of the new polystyrene beehives, the queen will lay right to the edge of the frame because she detects that the hive is better insulated and therefore warmer.

At the height of the season the number of bees will have increased to around 30,000 individuals, and as the queen continues to lay you have to be sure there is enough room. The queen must have sufficient drawn frame to continue to lay to maintain the growth of the colony. Otherwise, the queen might leave the hive to look for more comfortable accommodation. Make sure that there are frames to be made up in the brood box so that the colony can expand and check that once the super is full of honey there is another on the top to be drawn.

If the brood box is full and you are worried about the amount of space, you could add a super to the brood box without the queen excluder. This will give the colony extra space, which might be necessary if the weather is good and there is a plentiful flow of honey. In a disastrous year, on the other hand, you will have to feed your bees, even during the summer.

Rebuilding the hive

Return the frame that you laid outside the hive to its original position so that the box has its full complement of frames, then you can rebuild the hive, replacing each element in reverse order it was taken down and in such a way that you preserve the bee space. This means that you must position the queen excluder so that the 'gap' or bee space is the right way round – either at the top or bottom of the frames, depending on the type of hive. The same goes for the crown board, but you can get boards with a bee space on either side.

As you put the hive back together, for instance when positioning the super, excluder or another box on to the box containing bees, give them a little waft of smoke before carefully laying the box at an angle over the frame and then slowly twisting it into position. You are less likely to kill bees if you do this.

Take care too, as you are about to put a box with bees in it back on to a lower one, that you check that you do not have a pyramid of bees hanging from the uppermost box. To avoid crushing them as you lay down the box, gently shake them off and puff a little smoke to drive them away. Smoked bees usually climb downwards.

Paperwork

Don't forget to keep your notes about each hive up to date. A favourite place for keeping notes about a particular hive is in a plastic wallet attached to the top of the crown board. Keep a note of all the jobs you have done in the hive, such as adding supers, and whether you have seen the queen and eggs and brood. Keep a record, too, of any treatments you applied and the dates.

Above: *Always rebuild the hive completely and fasten up with the strap before starting work on the next hive.*

Checking frames

When you open a hive you should take the opportunity to check the frames. Before you remove any frames from the hive make sure that there are no bees on the top bar, especially where the top bar sits in the frame. Usually, the bees will have propolized the frame into position, so use your hive tool to loosen both sides of the frame before you lift it out.

Lifting the frame

Once you have removed any brace comb from the frame and have loosened it with your hive tool, carefully lift the frame from the box. It should be covered with bees, and they will begin to walk over your hands as they try to work out the change in their environment.

Below: Always lift the frame out of the box vertically and grab it by the lugs.

Examine both sides of the frame without loosening your grip on the top bar. This is an important skill to learn: do not be tempted to lay the frame anywhere other than in its correct position in the box. If you drop it you will be covered in a few hundred angry bees and run the risk of killing the queen should she be among them.

The base of the frames are frequently propolized, and you often find some comb drawn on the base. Clean this away with your hive tool and remove any wax on the top

Brace comb

Sometimes you will see what is called brace comb between frames. This is wax built by the bees between the frames to stick them together. If you look down vertically between the frames you will be able to see the comb, which is easily separated with a knife or hive tool. Brace comb contains no grubs, so you can break it with a clear conscience.

bar. If your hive tool meets some resistance on the under-side of the frame, do not force it. It might well be a nail or, worse, some of the wire inside the foundation. Work carefully because you do not want to tear open the grub-filled cells.

Queen cells

Swarming and re-queening hives are discussed on pages 84–7 and 88–9, but when you are carrying out a general inspection of the frames you might find queen cells in one or both of their typical positions.

If you find some queen cells in the middle of a frame then you need to check on the health of the queen. She might well be dead or not laying properly because these are emergency queen cells. You might also find queen cells that fall from the bottom of a frame and hang vertically; these are built in response to a failing queen or overcrowding.

If you know that your queen is young, fertile and healthy you should remove the queen cells. At the same time make sure that there is plenty of room in the hive for the queen to lay more eggs by adding another box. If you are uncertain about the age and status of the queen you should seek the help and advice of a more experienced beekeeper.

Egg cells

When you check a frame the presence of eggs will prove that the queen is working well. Studying the position of the eggs will enable you to work out how long ago the queen was last laying on any particular frame. Day-old eggs are glued vertically to the base of the cell. By the second day the eggs will have fallen over to an angle of 45 degrees, and on the third they lie flat on the bottom.

Recognizing a failing queen

A queen that is laying well will place her eggs in the warmest part of the frame, usually laying from the centre outwards. Her first working of a new frame will be in the centre, and these cells will be the first ones to be capped

Bee care

Checking a frame

While you are holding the frame look for the queen. Marking her (see pages 88–9) makes this easier.

If the queen is not on the frame shake down the majority of the bees from the frame into the brood box. A sharp, deliberate downward movement ending in a sudden upstroke of your hands should dislodge most of the bees.

Shaking down the bees will reveal the face of the frame. Look at the brood areas, with eggs and grubs and capped cells check the outer honey storage areas with the odd cell of pollen. If you see eggs you will know that the queen is working well.

Carefully lay each frame back into its original position or into a spare box beside the hive as you continue your inspection. Always replace the frames carefully to avoid killing any bees.

by the worker bees. In a circle around this you are most likely to see uncapped brood, and these cells will be surrounded by cells filled with stores, first pollen and then honey, which are set as arches around the central brood. (If you have a polystyrene hive remember that the queen will lay right up to the edge of the brood frame because of the extra insulation afforded by the hive walls.)

A failing queen will not lay in a uniform way. Some cells will have eggs in them that are not viable. The capped brood will then have 'holes' in it where the queen has missed. When the cells look patchy and weak the colony is in danger from the reduced numbers, and worker bees will set in motion the process of replacing the queen bee. This process is known as supersedure (see page 88).

Feeding bees

There are several times during the year when it will be necessary for you to give your bees a helping hand and provide them with food.

Why feed?

You should be prepared to feed your bees at any time of the year, and although there are three main periods when they are likely to need feeding, be ready with food during a cold, wet summer and whenever you have removed honey from the hive.

Above: *You can feed sugar syrup to your bees using a large bee feeder on top of the hive.*

Food is scarce

Winter is clearly a time when bee colonies may be hungry. Even though there are drastically fewer bees in the hive at this time, they consume huge quantities of honey: about 18 kilograms (40 lb). In winter each bee takes more than its own body weight of honey every couple of days.

In spring a young colony might not have sufficient flying bees to gather enough food from the environment, and feeding the hive with sugar will allow the colony to build up a brood. Creating wax requires large amounts of energy, not least because the wax itself has a very high calorie content.

Eventually, the working colony will find equilibrium between gathering bees and house bees so that food can be stored and the young tended, but at the beginning and end of the season the colony is vulnerable, and the bees have to be fed.

Autumn feeding

You will also need to feed your bees if you have taken honey or if there is not sufficient honey in the hive. It takes the bees a couple of weeks to create honey from sugar syrup and then cap it. The honey has to be matured by evaporation, otherwise it might well ferment in the cells and this would be disastrous.

You should only stop feeding the bees if they are no longer taking down the sugar or if winter sets in, which in practice means no later than mid-autumn. Make sure there is plenty of room for the bees to place the syrup into the cells.

Feeding honey

You should not give honey to your bees unless you are sure of its origin and disease status. Imported honey tends to contain nosema and foul brood spores, so be cautious.

Spring feeding

When the temperature begins to rise in spring, honeybees start to forage. This can be very early in the year, and often the natural supplies of nectar and pollen have not started to flow properly. Although bees rarely starve in winter, they are quite likely to starve in spring. Giving your bees some sugar syrup on warm spring days will allow the colony to make stores that will then enable them to keep pace with the expansion needed to take advantage of the late spring and early summer flow of available food.

Sugar syrup

The recommended standard for making sugar syrup is to use the ingredients in the proportions of 50:50.

Put 500 ml (17 fl oz) water in a large saucepan and heat it to boiling point. Turn off the heat and once the boiling stops stir in 1 kilogram (2 lb) granulated sugar. Keep stirring until the sugar has completely dissolved. When the liquid has cooled it should be clear. You can then transfer it to a bee feeder.

If you use more diluted syrup you will find the bees will need more time for it to evaporate before capping. Moreover, it might cause digestion problems and premature voiding of waste products because of the volume of liquid being taken by the bees. There is also some evidence that using unrefined or brown sugar in the syrup can induce dysentery in bees.

Because the syrup is odourless, you can make it more attractive to the bees by adding a little honey to the hot liquid. Make sure you add the honey when the water is very hot so that it kills any nosema spores or bacteria it may contain.

Do not put your finger into syrup to test the temperature. When sugar re-crystallizes the process gives off a lot of energy in the form of heat, and there is no burn worse than one caused by hot syrup, because as the liquid cools on the skin, sugar crystals continue to give off heat.

Sugar candy

Bees are not usually given candy, although it can occasionally be used in winter. Candy is used to make plugs for the bees to eat through – in a queen cage, for example (see page 57). You must provide water to help them get through the candy, otherwise they run out of spit to dissolve it.

Candy is made from a 5:2 mixture of sugar to boiling water. Once the sugar has dissolved, heat the mixture to 115°C (240°F) and pour it into moulds. It should set into a cake-like form.

Robber bees

Bees will rob each other of their stores, and this tendency can be exacerbated by making it obvious that there is sugar in the hive. The bees will give themselves away by flying around in an excited manner, and any bee that is already in the hive from another colony will soon communicate the presence of sugar syrup to its own hive. It is best to feed syrup towards dusk, thus reducing the likelihood of robbing, because bees tend to forget overnight. If robbing is a problem you should reduce the hive entrance so that your bees can defend it more easily. Take care that you do not spill syrup outside the hive because it will simply encourage others to explore the hive, particularly wasps, which will kill individual bees.

Bee care

Above: *A circular feeder is common and inexpensive and holds nearly a litre of syrup.*

Above: *This feeder has been placed directly on frames and the bees have glued it in position with propolis.*

Types of feeder

There are several types of feeder, and your choice is really a matter of personal preference or what is readily available.

Circular feeder

This is probably the most common form. It is cheap and easy to keep clean, but take care that you do not lose the central cup. The circular feeder holds about 1 litre (1¾ pints) of syrup. Inside there is an inverted plastic cup that restricts the bees' access to the liquid. Never use this type of feeder without a cup, and if you cannot find it use a honey jar instead.

The feeder can be placed on top of the crown board if it has holes in it, but it is actually better placed on top of the frames themselves so that bees can gain immediate access. It also cuts down the possibility of robbing by way of an ill-fitting lid. If you place this feeder on the crown board it will often fit under the lid, but if you put it directly on the frames you will need to put an empty box around it in order to close the hive. Of course, if it is placed on the crown board feeding hole then you will be able to fill it without disturbing the colony.

When you first put the feeder in position, dribble a little of the syrup down into the hive to gain the bees' attention and interest.

Contact feeders

This type of feeder, which is less popular than it used to be, is usually rested on the crown board. The jar is set upside down, and the lid allows the syrup through very slowly so that the bees can take the syrup. They are also able to speed up the flow by sucking hard.

Changes in the ambient temperature when the contact feeder is almost empty can cause the air inside the feeder to expand, causing a flood of syrup. You also need to keep an eye on it because once it is empty the bees tend to propolize it.

Frame feeders

This feeder is a box that is fitted in the brood box. It is the same size as a frame but is hollow, and there is a series of ledges and wire mesh inside to stop the bees from falling in and drowning. It is not suitable for winter feeding because you will need to open the brood box to fill it and this will chill the bees.

A frame feeder is the only effective method of feeding bees in top-bar hives.

Tray feeders

This type of feeder, which is used for the large-scale feeding of colonies, sits on top of the brood box. The tray will hold 2 litres (3½ pints) or more of syrup and allows bees access at one end. Tray feeders are built to fit into specific types of hive, so check that it is appropriate for your hive before you buy one.

How much food do bees need?

In order to get it through winter a strong colony will need about 15–18 kilograms (34–40 lb) of liquid feed. Each brood frame weighs about 2 kilograms (4¼ lb) when it is filled with capped honey, and a simple calculation will show that if all you have after the summer harvest of honey is a brood chamber full of capped honey this will hold about 22 kilograms (48 lb). But it is more than likely that the brood chamber is not completely full of honey at all. In addition, there will be an amount of pollen to take into consideration.

You will find that experienced beekeepers can lift the sides of the boxes (a process known as hefting) and know – apparently instinctively – how much food is available for their bees. You will do better to count the frames and work it out more accurately.

A super frame holds just over 1 kilogram (2 lb) of honey. You will therefore get 11 kilograms (24 lb) of honey in a full super, so a brood box and a super together should be enough to get a good colony through the winter. Check the weight of the hive by lifting and seeing how heavy it is.

Do not just assume that the colony has full brood frames: inspect them. You can then add the appropriate amount in late summer, remembering that there should be enough space for the bees to store all this feed and evaporate the water prior to capping.

Feeding pollen

Pollen is one of the bees' main sources of protein and fat, and it is also an important source of minerals. It has been calculated that to raise one bee the colony needs 100 mg of pollen. An average colony will gather around 50 kilograms (110 lb) of pollen to feed itself. Pollen is almost entirely used in rearing brood, but it is also important in stimulating the bees' hypopharyngeal glands.

The availability of pollen is not uniform, and in some countries it has been necessary to feed substitutes. Various materials, such as fish meal and soya flour mixed with sugar syrup, and sometimes a little egg yolk, have been used successfully. Most new beekeepers will not encounter this problem, and where it is necessary local beekeepers will have their own recipes that you should follow. Ultimately, you should try to grow plants that will produce a natural source of pollen.

Water

Make sure that your bees have access to water all year round. If you have a natural garden pond you will notice your bees in the shallow, warm water at the pond's edge. If you do not have a pond near the hive you must arrange a source of water.

Although you can buy some purpose-made water fountains, simple arrangements of gravel-, moss- or peat-filled dishes topped up with water are ideal. Set several of these dishes around the hive and in your garden.

Swarming

People are afraid of swarming bees for many reasons. If you are an urban beekeeper you are automatically in the spotlight if thousands of bees are flying around the neighbourhood and landing in gardens. But the primary reason beekeepers fear swarming is that the colony that remains is frequently weakened, leaving less honey to harvest.

Why bees swarm

Most of the books describe swarming as the bees' way of forming more colonies, but this is only part of the reason. Swarming is, in fact, part of the lifecycle of the honey-bee, and without it bees will not evolve further. As it is, the exchange of genetic material between bees takes place only every couple of years, and for this reason the honeybees' evolution is slow in comparison to predator insects and bacteria.

It is not possible to keep bees and make honey without mating a queen, and so the bees' sexual reproduction, during which the bees' characteristics are passed on to new generations, are part of the beekeeper's work. Fortunately for humans, subspecies of honeybees are genetically similar to each other, and so it is possible to hybridize bee populations to create strains of bee that have nearly all the qualities we would want, and nearly all the bees we keep around the world are hybridized to some extent.

There is sometimes no explanation of why bees swarm. You might well have the perfect colony, but the queen seems intent on vacating the hive. Should she repeatedly attempt this, it is probably advisable to re-queen altogether (see pages 88–9). This is a technical process for which a beginner may need to seek help.

Above: *A swarming colony alighting on to a cardboard box. You could also use a skep to collect the swarm, if you have one.*

Drones

Although they are often regarded as the unproductive members of the hive, drone bees are the driving force of a healthy bee gene pool. A hive may contain a few hundred drones, and each of these will have produced several hundred thousand sperms, each of them genetically different, and some perhaps with a beneficial mutation in the DNA that may make the bee a better gatherer, flyer, breeder, defender or survivor of varroa or another bee disease.

The numbers of drones will increase in spring as part of the natural routine of the hive, and at this time you should check the frames for queen cells, especially if your queen is greater than two years old.

Queen cells

The presence of queen cells in late spring is a sign that a colony is preparing to swarm. The two types of queen cells that may be seen (see page 79) might not be solely a response to an ageing queen. The colony might feel it is overcrowded; it might not have had enough stores to survive through winter without hardship; or the colony might be living through a prolonged period of bad weather.

Recognizing queen cells

You can recognize a queen cell by its unusual shape. It looks like a cased peanut protruding from the frame, either in the centre or at the bottom of the frame. Interestingly, the two types – centre or bottom – are built as a response to different stimuli.

The centre or supersedure cells are created in response to a failing queen. She might be old or ill – perhaps suffering from nosema or bee dysentery – and the supersedure cells are there to create a replacement queen. If you find them in your hive you may want to look closely at the queen's health.

The bottom or swarm cells are there to create a second queen to divide the colony, leading to more hives. It is a natural reproductive stimulus to create more bees.

Above: *When bees swarm, the queen is of prime importance and the workers fuss around to make sure she is neither lost nor damaged.*

Preventing swarming

If your queen is healthy and laying and is less than two years old you may wish to take steps to avoid the likelihood of swarming.

Remove queen cells

You can simply remove the cells with your hive tool. If there are no competing unmated queens your current queen is less likely to swarm.

Add a super

In mid-spring add a super to give the colony room to store the supplies that are now beginning to flood in.

Check the brood box

Make sure that there is plenty of space in the brood box for the queen to lay eggs. It is unlikely to be completely full in early spring, but you could add a super as a secondary brood box if you're really worried.

Above: *A queen marked on the thorax. She also has her left wing clipped.*

Clip her wings

You can prevent the colony from going more than a few feet by clipping one of the wings on the queen. If she cannot fly very far the rest of the bees will not wander far either. Be careful though, because queens that have had their wing clipped are often found on the floor in a swarm – a very vulnerable place to be.

Artificial swarm

If the queen is older than two years old you may consider making an artificial swarm. This is an easy technique to master and is usually successful. In the end you will have two colonies of bees instead of one, for nothing more than the cost of the boxes and frames needed for the new hive.

Move your hive so that it is about 4 metres (12–13 feet) away from its original position and set up a new hive in the old position. Put about seven brood frames in the brood box, preferably with a couple of them having already drawn comb so the queen can get an early start laying eggs. If necessary, introduce a frame of brood from the original hive that has got unfilled cells on it.

Some of the immature adults on that frame will be able to make an immediate start with undrawn frames.

Make sure there is space in the centre of the new brood box for a frame that you will transfer from the original hive. This frame must have the queen on it and no queen cells.

By this stage you will have a hive with a couple of frames of brood and a queen, and a hive with queen cells and a lot of bees and some brood waiting to emerge. You can remove the majority of the queen cells in the transferred hive. The first queen to emerge will mate and kill the others anyway.

All the flying bees that are in the transferred hive will naturally return to the hive that they recognize as home, so in the following hours the old queen will have most of her entourage back with her. The transferred hive will be queenless for up to a fortnight, but will accept the new queen once she has emerged.

It is a good idea to feed both the colonies of bees as a part of the artificial swarm process until they are freely foraging for nectar and pollen.

Introducing a real swarm

The swarm is a nightmare for beekeepers and something to be avoided at all costs among your own bees, but at the same time a swarm can be a gift when it appears – a colony for free.

Taking a swarm is not easy: you are, in effect, taking on someone else's bees. It is likely that the queen is older and has escaped for her life from a colony that has produced a viable new queen, and you have no way of assessing the qualities of the colony. Moreover, it is a distinct possibility that drones from the swarm colony could mate with new queens in your existing hives, thus introducing some of their traits into your stock.

You may find that your own bees swarm, and you are able to recover them yourself. This is especially true if your queen's wing is clipped because she will never get more than a few feet from the hive, so you will have to introduce your own swarm!

Collecting the swarm

Make sure that it is safe to recover the swarm – bees often settle high up in trees or on roofs. Then collect the swarm in a cardboard box or, if you have one, a skep (see page 44). Put the box or skep on a white sheet on the ground and give the swarm time to gather into the box. Some beekeepers bait the box with honey or even use pheromone sprays to attract the swarm, but this is rarely necessary.

When the bees have settled check for the presence of the queen, then transfer the swarm to the hive. If the swarm is small it may be possible simply to pour the bees into a brood box with the four centre frames missing but with at least a couple of drawn frames for the queen to inspect and begin to lay in. After a day you can inspect the swarm and close up the brood frames and fill the gap to complete the box.

If a fellow beekeeper can show that the swarm belongs to him you should allow him to take the bees back to their hive. Ask around before you finally secure them as your own.

Throwing a swarm

This is a job for two people and should not be attempted alone, particularly if you have never done it before. Attach a wooden platform to the entrance of the hive to form a ramp and secure it so that it cannot slip. Spread the sheet, with its box of bees, over the ramp so that some of the stragglers on the sheet can gain access to the hive, which can be baited with a little honey if required.

Take the box and carefully but firmly pour the bees from the box on to the sheet towards the entrance of the hive in a single movement. The word 'throw' is perhaps a little harsh: you do not want to kill the queen by throwing them with too much force. You should then see the bees start to walk along the cloth into the hive, with the bees at the entrance fanning their scent to those below. This encourages the rest of the bees to process towards their scent and as more bees enter the hive, the strength of the pheromones increases.

Bee care

Above: *Encourage as many bees as you can into the receptacle you are using.*

Above: *Gently throw the bees on to the sheet-covered ramp that leads to the entrance of the hive.*

Re-queening

The process of removing a queen and replacing her with a new one is called supersedure, and it should be done every two years. Re-queening ensures that the vigour and laying rate of the queen are as high as possible, and reduces the likelihood of swarming.

Natural supersedure

When a queen cell is allowed to develop into an adult virgin queen the process almost invariably leads to the old queen taking off in a swarm. This can be avoided by killing the old queen and removing all the queen cells but one. The new queen will emerge a week or so later, by which time the bees will be ready to accept her. She will mate and start her work.

Alternatively, you can buy or rear a replacement queen, introducing her to the hive in a queen cage (see page 57).

Supersedure should take place only when there are enough drones to inseminate the new queen, and this is usually the earlier part of winter. A barbaric-sounding method of forced supersedure is to clip off one of the hind legs of the old queen. She will not be able to place her eggs correctly, and the workers will realize this and start to prepare a new queen cell. They will then kill off the old queen by balling her, a process that involves a number of bees attaching themselves to the queen and increasing their body temperature until she dies from heat exhaustion.

Marking the queen

The life of the beekeeper is so much easier if the queen is clearly visible. She should, therefore, be marked. Putting a dab of white paint on her back will not only tell you where she is, but if you change the colour each year you can also easily tell how old she is. This isn't necessary for new beekeepers, who have only one hive, and nor is it essential if you keep proper notes in the hive. However, if you catch a swarm and the queen is marked you will immediately know quite a lot about the bees.

How to mark the queen

When you have identified the queen, catch her in your fingers and dab her with some quick-drying paint in the appropriate colour. This is not as easy as it sounds, and the procedure takes considerable experience. The queen is delicate, so you can easily damage her. Her legs are particularly important because she has to be able to

Colour coding for queens

The internationally agreed system of colour marking means that wherever you get your bees from or send them to will have the same system. Because a queen is rarely kept (or lives) for longer than five years the system repeats itself every five years.

- Year ending 0 or 5 – blue
- Year ending 1 or 6 – white
- Year ending 2 or 7 – yellow
- Year ending 3 or 8 – red
- Year ending 4 or 9 – green

squat into a cell to lay her eggs, and if she is injured she will not manage this.

The easiest way is to buy a queen cage, which is a trap that can be placed on top of the queen. You can mark her through the bars using a brush or matchstick coated in paint of the appropriate colour. Some beekeepers use special bee paints, while others use model-makers' paints.

Above: *A queen bee has been marked with yellow paint, indicating how old she is and making her clearly visible to the beekeeper.*

Moving a hive

There are several reasons for having to move a hive – you could be combining two colonies, for example, or moving your bees to a better food supply – but this is something that has to be done with great care and only after thorough preparation.

If you move the hive just over 1 metre (about 3 feet) the bees will congregate where the hive used to be and will take a long time to find it, sometimes too long, when the bees will die. Bees normally forage up to 3 kilometres (almost 2 miles) from the hive, and if the hive is moved within their familiar area they will fly back to where the hive used to be and will not find their new home.

If possible move the hive at dusk or dawn, when all the flying bees are inside the hive.

If you need to move a hive within the space of the apiary, simply move the hive a little less than a hive's width each day. Moving a hive further than this calls for rather more preparation.

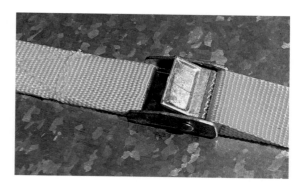

Above: *The clasp holds the strap in position – smear it with petroleum jelly to stop it rusting.*

Moving a greater distance

The hive must be secured so that the bees cannot escape in transit. The entrance to the hive must be stopped up and secured in position and the various boxes of the hive should be stapled together, or the hive should be securely strapped. Do not rely on carpet tape or any other single method. You need to be doubly sure the hive will stay intact and together.

Hives are heavy, and you will need help to carry the hives on to your transport. Use a proper levered truck to carry the hives, and never use a wheelbarrow because they tip easily and are difficult to control when they are carrying a heavy weight.

If you are moving bees in the height of summer you can remove the crown board and put a mesh screen in place so that there is increased ventilation.

Moving hives to and from various food sources is beyond the scope of the very new beekeeper, but bear in mind that once your colony has taken advantage of a food source, such as a field of oilseed rape or a heather moor, there will be a lot more bees in the hive than when you originally moved it. Retaining all of these bees might need the application of a fresh super full of foundation.

Moving bees in nucleus boxes

Perhaps the most convenient way of moving a single hive is to put the frames at dusk into nucleus boxes. These boxes normally have a deep, fitted lid, so there is no need to over-secure them. The rest of the hive can then be dismantled and transported safely.

Transport

When you move your hives by road the journey should be as smooth as possible. Plan your route in advance and leave plenty of time. Throwing the colonies around inside the hives can cause them to abscond when they get a chance to escape and, of course, you need to be sure that you do not damage the queen.

Above: When moving your hive, ensure that you use straps to keep it secure.

Pests and diseases

Bees are susceptible to an array of pests and diseases, but fortunately not all of them are devastating to the colony. Indeed, until the appearance of the varroa mite it was only acarine and foul brood, in its various forms, that were really deadly. Now a number of new problems beset honeybees.

Be aware that American foul brood, European foul brood and the pests, small hive beetle and tropilaelaps mite, are notifiable in most countries. This means that if you have, or suspect you have, any of these problems you must contact the appropriate inspector.

Hygiene

Most diseases get into hives because bees drift from one hive to another or because robbing bees and pests have entered a hive. It is also possible that beekeepers may introduce disease by failing to observe high standards of cleanliness when they are around and inspecting the hive. Hygiene is particularly important for people who regularly visit more than one site, perhaps members of an association or people who have apiaries in several places.

If you are dealing with more than one hive wear disposable gloves and change them between hives. Always make sure you disinfect your hive tool between hives. You can keep a solution of hypochlorite for this purpose. Make sure you always clear away debris from around the hive. Do not leave any cappings or end-of-frame cells that you remove around the hive, because they only invite robbing and other pests.

When you use old items in a hive – boxes, floors queen excluders and crown boards – everything that the bees can touch should be flamed with a blowtorch before use. Old wax should be rendered and the frames replaced. Plastic hives should be cleaned with hypochlorite.

Pests

Bees are much more susceptible to attacks from a range of pests during the winter months when they are not actively guarding the hive.

Some pests are more surprising. In winter woodpeckers are capable of making large holes in hives, but can be deterred by covering the hive with a wire mesh so that bees can fly in and out.

Robbing bees and wasps can be a problem, especially if you have recently fed the hive (see page 81).

Varroa

Varroosis is the most serious problem facing beekeepers. The mite *Varroa destructor* causes deformities in bees and will eventually lead to the collapse of the entire colony if it is untreated. The red, pill-shaped mites are just over a millimetre in diameter. It was originally a pest of the Asian honeybee, *Apis cerana*, and these populations seem to have built up some immunity. The Western honeybee, *Apis mellifera*, however, has none.

Although it is serious, varroa is not a notifiable disease because it is ubiquitous – everyone has it and every beekeeper has a responsibility to keep the problem under control. Infestations are now found all over the world, except Australia, and within a decade it has become an important apine pandemic. Sometimes a few thousand mites can be fatal to a colony, but in other hives only a few hundred mites can wipe out the bees.

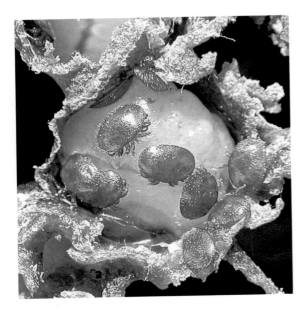

Above: *Varroa mites sucking the life from a developing grub.*

The numbers of mites fall in winter – they are unable to grow in brood cells because the queen has usually stopped laying – but the remaining mites huddle with the bees in their winter ball and live off the haemolymph of adult bees.

Pyrethroid insecticide treatment

For several years a pyrethroid insecticide was used to control varroa, and beekeepers kept treated strips in the hive. It did not take long for the mites to become immune to insecticides, and this kind of treatment has been superseded. A thymol-based product is now used, and it works in such a way as to make it almost impossible for mites to become immune (see page 94).

The problems that pyrethroid resistance brought has led to the concept of managing bees through an integrated pest management system. The basis of the system is knowing how many pests you have in your hive and the likely outcomes of their presence at particular times of the year.

Developing an integrated system

You must first make a judgement about the health of your bees, how the colony is doing and how they are likely to do in the future. Then you can implement a couple of basic procedures to reduce the number of mites in the colony without having to resort to chemicals.

First, there is a lot of evidence that mites that fall off the frames alive are able to climb back into the brood box if they hit a solid floor. Therefore, if you put a mesh floor in the hive the mites are more likely to fall right out of the hive and find it impossible to get back into it. It is possible to account for up to 30 per cent of the varroa population in this way.

Second, the female mites prefer to feed in male, drone brood. You can encourage drone brood production by putting a super frame into the brood box. When the worker bees make hanging comb in this formation it is usually drone brood. By simply removing the brood when it is capped you will also remove a lot of mites.

Much depends on the overall health of the bees, and the presence of tracheal mites is also thought to be an important factor.

In an infected hive the brood emerge with deformities that make them useless to the colonies. Others die at the pupal stage in the brood itself and remain until the adults remove them. The beekeeper will notice a patchwork of dead brood, and if nothing is done the colony will be in danger of collapse.

The mites are spread over huge distances on the backs of bees, and hives are infected by the natural processes of swarming, and robbing. The adult female mites will enter a brood cell just before it is capped and will feed on the haemolymph of the immature bee. She will also lay eggs in the cell and the offspring mate in the cell. When the bee emerges from the cell the mites escape, although only adult females survive – immature mites and males are not viable.

The mite count

Clearly, given the ecology of the varroa mite, the numbers will increase in proportion to brood production, and from late spring the mite numbers will follow the activity of the hive. You need some way of working out when to intervene with harsher remedies.

The increase in mite numbers depends primarily on the population of mites at the beginning of the season. If there are 100 mites in the colony in spring you are likely to see over 1,000 by summer, and although it is difficult to be sure and exact in these circumstances, a hive population of 1,000 mites seems to be a critical level.

It is possible to estimate the number of mites in a colony either counting the number of dead mites falling from the frames through the mesh floor and on to a varroa floor beneath it or by looking at the number of capped brood cells that have mites inside them. Monitor your bees at least four times during the year: in early spring, after the spring honey flow, at harvest time and again in the autumn. The bee club will have a microscopist to help.

If you collect mites over a period of, say, five days you will be able to determine a daily mite fall rate. It has been estimated that a colony is in danger if the fall rates are around: 0.5 mites in early spring; 6 mites in late spring; 10 mites in early summer; 16 mites in midsummer; 33 mites in late summer; and 20 mites in early autumn.

The other method involves uncapping about 100 brood cells that are just about ready to come out. If you find that about 10 per cent of the cells contain varroa, the colony is in trouble.

Other treatments

Now that varroa mites have developed resistance to pyrethroid insecticides, other methods have to be used to minimize mite numbers. Beekeepers should take care that insecticide residues are not found in honey. Research into new treatments and special breeding programmes to create varroa-immune bees are under way.

The available pesticides still include some pyrethroid treatments, as well as aromatic products based on

Above: *Wasps robbing honey from a frame.*

thymol. Follow the manufacturer's guidelines to the letter, especially as regards the timing of applications.

Another, non-approved treatment that is popular with many beekeepers uses oxalic acid, which is simply dribbled between the frames as a 3 per cent solution.

Mice

In summer, when the bees are active, mice will usually be deterred from entering the hive by the stings, but in winter, when the bees are not active, mice can be a real nuisance if they get into a hive, making their nests from grasses and leaves and chewing combs. Moreover, once a mouse is actually in the hive its offensive smell deters bees from attacking it. Keeping mice out in the first place is best, so fit a mouse guard, a piece of metal (usually zinc) with holes in it, over the entrance in late summer to early autumn and remove it in spring, when the bees are fully active. In winter, on the days when it is warm enough for bees to fly, they will be able to get through the holes in the mouse guard.

Wasps

In late summer and early autumn wasps often try to enter the hive to take honey. Keep them out of the hive by reducing the size of the hive entrance – it is easier for the guard bees to protect a small entrance. Make sure that the outside of the hive fits perfectly together so that there are no small cracks through which a wasp can enter to rob the honey.

Wax moths

Two kinds of wax moth, the greater and lesser, are more of a problem to stored comb, which they eat, than to the bees themselves. The moths are rarely visible in summer when bees are active, and a healthy colony will often be able to expel any intruders. In autumn and winter, however, the moths tunnel into wax combs, especially brood comb, to lay eggs and produce silvery threads. The larvae of the greater wax moth also chew indentations in wood (both in the hive itself and in stored hive furniture) where they spin cocoons in preparation for pupating. The bees seem to leave them alone.

One of the easiest ways of treating stored frames is to put them in the freezer for 24 hours or so because the larvae are susceptible to cold. In the hive you will have to use a proprietary product.

Diseases

The most important disease facing bees is varroosis, the disease caused by the varroa mite, which is discussed earlier, but there are several other diseases that can be troublesome.

Foul brood

Both types of foul brood, American and European, are notifiable diseases. They are caused by bacteria. American foul brood (AFB) causes infected bee larvae to die in the cells, while European foul brood (EFB) tends to affect unsealed brood. There is no known treatment for AFB, and both the hive and its contents must be destroyed if the disease is discovered. Minor outbreaks of EFB can sometimes be overcome by shaking the colony into a new hive, but if this is ineffective, again, the hive and contents must be destroyed.

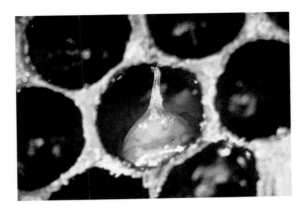

Above: *The stringy form associated with foul brood.*

Nosema

A disease of the bees' digestive systems, nosema is caused by a protozoan. The bees suffer from dysentery, and the infection is spread when workers in the hive try to clean up the mess by licking it. Nosema weakens bees, preventing them from functioning as normal, but it is rarely fatal. Infected bees can be treated with antibiotics given in sugar syrup in autumn. Avoid infection by maintaining scrupulous hygiene in and around your hive and by thoroughly cleaning all stored hive furniture and by fumigating empty hives.

Acarine

The disease is caused by a mite, and it affects the bees' trachea (breathing tubes). Although it is not fatal, bees are weakened and severe infestations of the mite can lead to the collapse of the entire colony. There is no known treatment, but a proprietary menthol gel has been shown to have some effect. Some bee hybrids seem to be more susceptible than others.

Honey and other products

Many people develop an interest in beekeeping because they want to enjoy the honey that is harvested in late summer, and from a well-managed hive it is possible to get up to one hundred jars. Honey can be used in dozens of recipes, some of which are given on pages 124–35, and in mead. Bees also produce wax, which can be turned into candle or polish, and propolis is also collected.

Honey

For thousands of years the goal of our relationship with bees has been to remove the honey from the hive. A well-managed hive can produce far in excess of the bees' requirements, leaving the beekeeper free to harvest it.

Harvesting honey

Remember that when you harvest honey you should replace it with an equivalent quantity of sugar syrup. A careful beekeeper will not take more than the colony can easily afford to miss. Putting the bees in danger for a few frames of honey after a cold, wet summer is counterproductive.

The honey you take will be probably be in the topmost supers, and you should leave one full super for the bee's winter store. You can remove one super to make room for a feeder, set on a crown board (see page 82), and this will allow you to top up the syrup quickly and without chilling the bees on cool autumn days.

Expelling the bees

To get the honey from a super you need to find a way of removing the bees. If you set the super on the crown board with bee escape valves in the holes, the bees can go down into the colony but not return. Put the roof on and wait for at least a day before opening the hive again. Do not wait around when taking honey.

Remove the roof and take the super well away from the hive. Remember to take out the bee escape valves and set the feeder in place over the holes in the crown board. Put the roof back on. This method has minimal impact on the bees.

There will still be a few bees on the frames, so remove them carefully so that they can return to the hive. Putting the frames inside a sealable plastic box will allow you to remove the honey from the hive area.

Above: *Using a hot, sharp knife remove the wax cappings, using the frame as a guide.*

Getting honey from a frame

Removing honey is a complicated business: you have to remove the cappings, extract the honey and finally transfer it to jars to store it. You will need honey knives, extractors, filters and sterile jars all ready and available at the right time and in the right place. If you can, extract your honey at your local beekeepers' club or association, but if you cannot, make sure that all the surfaces and equipment are sterile and that you can work without being disturbed. Close the doors and window so that bees or, worse, wasps aren't attracted to the honey and make a nuisance of themselves.

Moving supers

A super full of honey is surprisingly heavy, and if you have more than one to carry you should use a trolley. Always use both hands to collect and move honey, and make sure that when you lift you do not strain your back. If you find it difficult to move heavy weights, ask someone to help you.

Above: *Decapped honeycomb dripping on to a tray.*

The jars used to store and sell your honey are often traditional, and most hold the metric equivalent of 1 lb (455 g). Be wary of using washed jars from last year, particularly if they have been used for anything else. Honey absorbs flavours and smells very easily, often making them more intense. Pickle vinegar is probably the worst offender. You must use sealable jars because honey is hygroscopic – it will draw water from the atmosphere – and this can lead to fermentation and consequent spoiling. You can usually buy special honey jars quite cheaply from your local bee club or beekeeping stockist.

Removing the cappings

Use a capping knife – an electrically heated knife that slices through the wax – to open the cells. Stand the frame on its side in a container in which you can collect the cappings, then cut away the capping on both sides of the frame (the honey will start to dribble out) and put the frame into the extractor.

If you do not have a heated capping knife you can buy a comb decapper, which resembles an ordinary hair comb but has angled metal splines. You can also use a serrated kitchen knife, heating it in a container of boiling water. Make sure that the container you use for the hot water is completely free from taint and residues. Save the cappings: they have honey on them and the wax is valuable and can be processed later (see pages 104–5).

Extracting honey

If you do not have an extractor, rest the frame over a large, sterile, taint-free, food-grade container or bucket. Gravity will cause the honey to dribble from the decapped combs and into the bucket. This takes ages to complete and you will never get all the honey out of the comb, so it is worth using an extractor.

Honey extractors rely on centrifugal force to remove the honey from the frames. The frames are inserted into a drum, which spins, slowly at first, so that the honey is thrown out of the cells on to the inner casing of the drum, from where it falls into the reservoir below. You might have to repeat the spinning process after moving the frames around to make sure you get all the honey from both sides of the frame.

A tap at the bottom of the reservoir has a coarse filter to remove any large pieces of wax. From here the honey is transferred to a holding vessel for 24 hours to allow the bubbles to rise to the surface. A second filtering gives especially clear honey.

Leaving the honey in the settling vessel so that air bubbles can rise to the surface does mean that the honey will have cooled, and this will make second filtering and pouring into jars difficult.

How to harvest honey

1 Use a comb or a knife to decap the cells in the honey frames. The decapping comb is more efficient.

2 Load the extractor so all the spaces are filled. Make sure the drum is evenly balanced.

3 When the extractor is full, start the motor (or turn the handle) slowly so it does not shake, and gradually increase the speed to maximum. The extraction process can take 10–20 minutes, and then you should check both sides of a frame for residual honey.

4 Replace the frames, once extracted, into the super box and put this back on the hive for the bees to reclaim the last droplets of honey for their own stores.

5 Once the honey has rested to remove the air bubbles you can pour it off into a food-grade bucket or honey jars. This takes a long time and can be quite sticky.

6 The decappings will be full of honey. Store them in a colander or a warm gravity extractor to get the last drops of honey from them and finally use the remaining wax by washing it thoroughly in plenty of water to remove any remaining sweetness.

Honey and other products

Types of honey

Whether the honey is runny (clear) or set honey depends on where the bees have collected their nectar.

Set honey usually depends on the crystal size, and honey collected by bees that have visited oilseed rape sets very hard. All set honey can, of course, be melted by gentle heating. Do this in a water bath or bain marie so that the combs are not subjected to direct heat. If you are able to remove the frames from the hive as a soon as they are full the honey will not granulate immediately and will consequently be easier to extract. However, all honey granulates after a time, a phenomenon caused by the relative concentrations of glucose and fructose.

It is possible to blend honeys from various sources to create a range of flavours, but the actual nutritional value of the honey will be unaltered. Honeys from single plants generally have interesting fragrances and flavours, but when you consider how far bees will have travelled to forage, you can see that your hives would have to be set in a huge area devoted to a single plant – such as a heather moor – to give a single floral honey. For the same reason it is difficult to obtain truly organic honey.

Dark honey

Heather is a dark, richly flavoured honey, which is gathered from moorland in late summer. Heather honey is somewhat gelatinous and will not come out of the cells under normal extraction methods. The combs are removed and pressed in muslin cloths so that the honey is forced out of the comb under pressure.

Cut comb honey

Cut comb honey is created either with unwired foundation or in a top-bar hive, where the bees create all the comb without the aid of foundation.

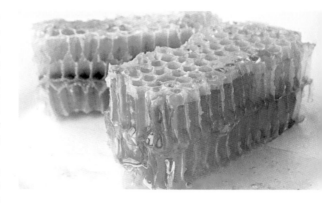

Above: *Comb honey is delicious and attracts a premium in the shops.*

The comb is not uncapped, and the honey remains sealed and in pristine condition. This honey attracts a premium price among those who like it. If you have a top-bar hive you can simply cut the honeycomb.

Selling honey

If you are simply selling a few pots of honey to friends and family you can usually do so quite freely. However, you must be aware of the legislation and regulations governing food production, farming and labelling, and you must make sure that your honey meets specified standards if you plan to sell a significant amount.

Propolis

Also known as bee glue, propolis is a mixture of various resins
and wax gathered by the worker bees from plants.

Propolis in the hive

This substance is filled with terpene and other aromatic
chemicals, and it is used by bees within the hive.

Its main function is to 'glue up' the hive, which
improves its structural integrity, but bees also use it to
propolize almost anything that takes their fancy, which
can be irritating for the beekeeper but is useful to the
bees (see page 33). It used to be believed that it was
used to reduce draughts by filling all cracks and gaps,
but this theory has been questioned in recent years as
research has shown that bees benefit from being in a
well-ventilated hive. Nevertheless, most beekeepers will
leave propolis in place in winter. It is also reduces vibra-
tions in the hive and makes the hive easier to defend by
narrowing gaps.

Propolis is also used as an external antiseptic. It fights
infection in the hive remarkably well, and this has led to
the idea that it can be used in human medicine.

Propolis and humans

Chinese and homeopathic medicines use propolis to
treat burns, ulcers, allergies and infections, and although
much of the evidence is anecdotal, there is some
evidence that it can be successful.

It has been used to treat viral infections, particularly
sore throats, and to this end is made into a tincture.
When propolis is dry it hardens almost like rock, and
many beekeepers pass a small lump around their mouths
as a guard against colds. They say it works, although a
proper scientific investigation into these claims has yet to
be undertaken. Chemical analysis has revealed that
propolis is high in antioxidants and that it has anti-fungal
and some anti-viral properties, but scientific tests have
not yet confirmed these findings in humans.

It is deeply coloured, and if you use propolis it will
stain anything it comes into contact with and will resist
any attempt to wash out. It is said to have been used for
many hundreds of years as a colouring for musical
instruments.

You can buy or make your own propolis dissolved in
alcohol, when some of its efficacy is said to be removed.
Although it is harder to dissolve, propolis in water retains
most of its active ingredients. It is not likely that you will
be able to sell your propolis for medical reasons, but you
can use it yourself.

Below: *Propolis looks unappetizing, but is used because of its health
promoting properties.*

Honey and other products

Beeswax

Beeswax is a remarkable substance that is produced by a series of glands on the underside of worker bees and is then manipulated by the various house bees to create the amazing architecture of the hive.

Using beeswax

The production of beeswax was once a major industry, particularly for the making of candles and polish. Beeswax candles, in fact, remain one of the only truly carbon-neutral sources of light because all the carbon dioxide given off by the candle has been taken from the atmosphere by the plants that made the bee food.

When you are extracting honey (see pages 98–101) the excess honey that is attached to cappings and comb can be extracted further by wrapping them in muslin and allowing them to drain. From there they can be soaked in boiled, cool water to dissolve the honey (this liquid can then be used to make mead, see page 134). The wax can be allowed to dry and prepared for use.

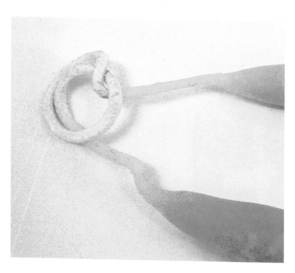

Melting wax
Do not overheat wax. The best method is to use a heatproof dish set over hot water – in the same way you would melt chocolate – or a double saucepan or bain marie. Do not boil wax, because it can flame.

One method is to put the wax in a muslin bag and immerse this in water. Gently heat the water to about 62°C (144°F), when the wax will melt and rise to the surface of the pan.

Alternatively, you can use a solar extractor, which is a glass frame that is orientated towards the sun, until the temperature reaches 60°C (140°F). The container is made of metal and has a window through which the sunlight enters. Put the wax on a plate set at an angle of 45 degrees. Sometimes the plate rests on metal to increase the heat retention in the extractor. As the wax melts it flows on to a collecting plate on the floor. You can use this wax to make new foundation or sell it, and some companies will allow you to barter wax for foundation.

You can also make the wax into candles, which is more difficult than it might seem because you have to make sure that the wick goes down the centre. You can buy moulds to make professional-looking candles.

Left: *A set of home-made beeswax candles.*

Beeswax products

Before using beeswax cosmetics on the sensitive skin of your face and neck, always check that they are safe by testing them on your hands.

There are lots of recipes available so that you can make your own soap, lip balm, make-up remover and a host of other products. By way of illustration this hand cream recipe is excellent for those working outside and the lip balm is the richest you will ever find.

Basic hand cream
This could hardly be easier to make. In a double boiler melt together equal amounts of beeswax, cocoa butter and almond oil. Stir thoroughly to combine, and then pour the mixture into a mould. If you wish you can add a few drops of essential oil.

These make delightful gifts. For instance, if you add a few drops of lavender essential oil you could decorate the bar with a sprig of lavender flowers added to the mould. The cream smells wonderful and is antiseptic.

Lip balm
Mix together 100 g (3½ oz) coconut oil (or almond oil or a mixture of the two) and 60 g (2½ oz) beeswax. Melt gently in a double boiler and blend thoroughly, then transfer to a small pot for ease of use.

You can add all sorts of things to this basic recipe, including a vitamin E tablet and a drop of essential oil (try peppermint or rosemary), or you could change the oils and use light essential oils or shea butter instead.

Furniture polish
This wonderful substance is simply a mixture of two parts pure turpentine to one part beeswax. You have to be very careful when you're making it to avoid fire. Always wear gloves and observe proper fire precautions.

Carefully heat 500 g (1 lb) wax in a tin set above a saucepan of water. Make sure that you keep the turpentine well away from the heat source. When the wax is completely melted turn off the heat and take it away to a separate place, far removed from any flame. Pour it into a large tin with 1 litre (1¾ pints) pure turpentine and stir well to combine thoroughly. Pour the warm mixture into moulds and when it has set keep it in a dry, cold place. The mixture is flammable, so always handle it with care, but it is the very best polish.

Right: *Beeswax can be used to make a variety of cosmetic and household products, from lip balm to furniture polish.*

The bee garden

The sound of bees as they buzz from flower to flower gathering pollen is one of the pleasures of the summer garden. However, beekeepers will be anxious to ensure that there are sources of food for their bees from early spring through to autumn, whether this involves siting the hives near to suitable plants or providing as wide a range of plants as possible in their own gardens.

Bees and biodiversity

Botanists have long studied the rise of flowering plants (angiosperms) but have not always appreciated the remarkable part that bees have played in this story.

Bees and pollination

The honeybee and its cousin, the bumblebee, are ideally adapted to the job of pollination and are by far the most important pollinators and consumers of nectar. Even if there were no humans, honeybees would still be servicing the huge range of plants and flowers that cover the earth. Bees and plants exist only for their own purposes, and it has to be said that humans represent the greatest threat to bees, largely from pollution.

Most pests and diseases have evolved without the aim of eradicating their host, yet humans have endangered bees around the world, partly through the destruction of their natural habitats, partly through pollution and partly by moving bees from country to country so that new diseases and problems arise so quickly that the bee has little time to find its own solutions and defences.

In the garden, what is good for the bees is also good for other insects – butterflies, moths, hoverflies and even wasps – and every living creature has a place in the complex ecosystem that exists in every back garden. These days most gardens contain conglomerations of plants from all over the world that would never have grown together in the wild, and it is up to the gardener to make this work to the benefit of the insects.

Left: *Bees have evolved to draw nectar and pollen from flowers and nothing else.*

Bumblebee nest box

Bumblebees are very efficient pollinators and are worthy of encouragement to the garden, particularly as they often struggle to find natural nesting sites. To help the population flourish, try making a homemade bumblebee nest box in the early spring.

Take two plastic flowerpots, line with moss and secure the open ends together with watertight sticky tape. Site the box in a sheltered spot – under a hedge is ideal – by burying one end in the ground with the drain holes above the surface, through which the bees will enter. Alternatively, position the nest box under a shed.

Right: *Make your apiary as beautiful as possible; more flowers mean more honey.*

The bee garden

Gardening for bees

There are few things more enjoyable for the beekeeper than being able to watch his own bees foraging in his own garden and then enjoying a crop that other people will miss: the millions of droplets of nectar, condensed and preserved and converted into wonderful honey, full of the goodness of the garden.

Above: *The aroma of plants and flowers is very important to foraging bees.*

Choosing plants

In the garden, as in nature, plants respond to being pollinated by bees and go to extraordinary lengths to avoid passing their own pollen to their sex cells. It is known that the yield from crops is improved by 30 per cent if bees are available for garden pollination.

The major sources of pollen and nectar include fruit trees, especially *Malus* (apple) and *Prunus* (cherry),

Salix spp. (willow), *Tilia* spp. (lime) and *Trifolium* spp. (clover). Bees also visit *Brassica napus* (oilseed rape) and *Erica* , *Calluna* and *Daboecia* spp. (heather, ling). As well as these, your garden can include a wealth of other pollen-rich plants, from *Lavandula* (lavender) and *Buddleja* (buddleia) to roses and *Lonicera* spp. (honeysuckle). There should be flowers in every month.

Take the time to cultivate your plants as naturally as possible, and this includes your choice of species and cultivars. Modern plants have been hybridized and bred for their colour, habit of growth, keeping qualities and disease resistance, and it is usually the older, traditional varieties and cultivars that are richer in pollen and nectar. Also try growing a packet of wildflower seeds.

Beekeepers will avoid growing double-blooming flowers, which produce much less nectar than single-blooming types. It is also more difficult for the bees to get to the nectar and pollen of flowers that have multiple petals, which can actually exclude bees and other pollinating insects. Also avoid F1 hybrid plants, many of which have impaired pollen production because they have been hybridized so that they will not set seed.

Techniques

Needless to say, you should practise organic gardening, avoiding the use of all insecticides and sprays and allowing your garden to develop a natural balance of pests and predators.

Above: *Medicinal plants, like this lavender, do pass on some of their flavour and properties to the honey.*

One useful technique that is often overlooked is dead-heading. Regularly removing pollinated flowers cuts off the hormone supply of the ripening seeds in the fruit or seedhead. These hormones suppress flower production, and if you deadhead you are more likely to get extra blooms on each plant and, consequently, more food for your bees.

Get into the habit of feeding your plants with good-quality, well-rotted compost. Not only will regular applications improve the soil quality and structure (giving healthier plants), but the phosphates and nitrates it contains will encourage your plants to produce better quality pollen and nectar. Pollen in particular needs nitrogenous excess, and the more there is available to the plant the more nutritious the pollen becomes.

Keep your plants well watered so that they do not suffer from water stress. Wilting plants shut down photo-synthesis, and this will be reflected in nectar flow. If possible, use collected rainwater and always water at the roots rather than the leaves.

Remember that by surrounding your garden with high walls, hedges or fences your bees will be forced to fly high and will not bother the neighbours.

Best plants for early nectar

In addition to the ornamental plants noted here, fruit trees, such as *Malus* (apple), *Prunus* (cherry) and *Pyrus* (pear), produce blossom in early spring and depend on early-flying insects, including bees, to pollinate them. If you have space, a couple of flowering cherries will keep the hive busy for a week.

- *Anemone nemorosa* (wood anemone)
- *Aubrieta* (aubretia)
- *Aurinia saxatalis* (gold dust, yellow alyssum)
- *Cyclamen coum* (cyclamen)
- *Eranthis hyemalis* (winter aconite)
- *Galanthus nivalis* (snowdrop)
- *Hyacinthoides non-scripta* (bluebell)
- *Muscari botryoides* (grape hyacinth)
- *Primula veris* (cowslip) and *P. vulgaris* (primrose)
- *Ribes sanguineum* (flowering currant)
- *Viola odorata* (garden violet, sweet violet)

Best plants for late nectar

All the plants listed here will flower from midsummer and, depending on the weather, into autumn. Deadhead regularly to encourage the plants to produce new blooms.

- *Aster* spp. and cvs. (aster, Michaelmas daisy)
- *Calendula* spp. (pot marigold, English marigold)
- *Centranthus ruber* (red valerian)
- *Echinacea purpurea* (coneflower)
- *Hedera helix* (ivy)
- *Helianthus annuus* (sunflower)
- *Lonicera* spp. and cvs. (honeysuckle)
- *Sedum spectabile* (ice plant)
- *Solidago* cvs. (golden rod)
- *Tagetes* cvs. (marigold)

The bee garden

Above: *You can increase the production of new flowers by deadheading old ones.*

Planting

With some planning and forethought it is possible to have flowers available at most times of the year when bees are flying, from early spring to late autumn.

The salvation of many bee colonies, because of the huge amounts of pollen it produces, is the plant *Impatiens glandulifera* (Himalayan balsam), a tall annual, bearing white, pinkish or red flowers from midsummer to early autumn. It is however, invasive, explosively spraying seed for metres around each parent plant, and in some areas, where it has escaped from gardens into the wild, it is regarded as a pest and must be uprooted wherever it is found. Beekeepers, however, are delighted when they see their worker bees landing on the alighting board covered in pollen as though they have just escaped from a box of caster sugar.

Other plants that used to be called weeds have become accepted in the garden. Buddleias, named after the Rev. Buddle, still found growing wild on rubble and other inhospitable places, have found a new niche as the butterfly bush. They are equally palatable to bees, however, and cultivated varieties, in shades of white, blue and purple, are just as useful as the species. Other plants, such as fragrant *Lonicera* (honeysuckle), are among the best providers of energy for your bees.

Plants for honey flow

Early spring
- *Alnus* spp. (alder)
- *Populus* spp. (poplar)
- *Prunus avium* (wild cherry) and *P. spinosa* (blackthorn, sloe)
- *Salix* spp. (willow)

Mid-spring
- *Malus* spp. and cvs. (apple)
- *Prunus* spp. and cvs. (cherry, plum, etc.)
- *Pyrus* spp. and cvs. (pear)
- *Taraxacum officinale* (dandelion)

Late spring
- *Aesculus hippocastanum* (horse chestnut)
- *Crataegus* spp. (hawthorn)
- *Rubus idaeus* and similar (raspberry, etc.)

Early summer
- *Brassica napus* (oilseed rape)
- *Tilia* spp. (lime)
- *Trifolium* spp. (clover)

Midsummer
- *Filipendula ulmaria* (meadowsweet)
- *Lupinus* spp. and cvs. (lupin)
- *Rubus ulmifolius* (blackberry, brambles)

Late summer to early autumn
- *Calluna, Daboecia* and *Erica* spp. (heather, ling)

Useful midseason plants

Perhaps surprisingly, midsummer is often a time when flowering plants are in short supply. After the burst of spring-flowering bulbs and shrubs there is often a lull, and beekeepers need to plan their planting carefully so that there is a constant supply of nectar and pollen.

Although they are widely regarded as a weed and are dug out and cut back in most gardens, *Rubus ulmifolius* (brambles) are a great source of nectar for the bees. They are in flower from early summer and are never free of bees.

Lavandula (lavender) produces huge amounts of nectar throughout the summer, especially when there is plenty of nitrogenous material in the soil.

There are several types of spiraea, medium-sized shrubs that bear milky pink and white flowers. They are all prolific pollen producers full of nectar, so it is not unusual to find bees on the flowers all day long.

Buddleias can be cut back hard in early spring, delaying the flowering until mid- to late summer. Cut back the stems to about 45 cm (18 inches) high, and it will grow to about 2 metres (6 feet) high and produce a mass of nectar-rich flowers. If you have more than one buddleia, prune them at different times so that you have flowers from early summer to mid-autumn.

Flowering onions, *Allium* spp. and cvs., produce masses of nectar and are also interesting to look at. They bloom from early summer onwards, and you will see bees clustering around the flowerheads. The flowers of the related *Allium schoenoprasum* (the herb chives) are also rich in nectar as well as being useful in the kitchen. These plants are a valuable source of the chemical sulphonamide for the colony, because they are highly antibiotic and important for the health of the hive.

All the labiates (nettles, dead nettles, vetches, peas, beans, lupins and so on) seem to have evolved with bees in mind. Some of the flowers can only be accessed by bees, and the blooms are a useful source of good-quality nectar and pollen.

The wildlife garden

Ideally, every garden should have an area that is set aside for wildlife, but that is not always possible in a small garden.

However, if you have room for a section of your garden to be devoted to wild plants, this will be a tremendous boost for your bees, because most produce fantastic amounts of pollen and nectar.

Above: *Create a feast for bees consisting of as many different flower types as possible.*

One of the bees' favourite wildflowers is the pink-flowered *Chamerion angustifolium* (rosebay willow herb), which is especially rich in pollen. *Convolvulus* spp. (bindweed), which can be such a pest in the tidy garden, is also a great producer of pollen, as is *Taraxacum officinale* (dandelion) in spring. Different species of *Trifolium* (clover), now banished from lawns, were once a major food for bees and were widely grown as a green manure.

Many herbs are attractive to bees, and a wildlife garden is a great place to grow *Melissa officinalis* (lemon balm), whose Latin name suggests that it is the official bee plant. It is also known as bee balm and will grow well in most reasonably well-drained soils, but take care because, like *Mentha* (mint), it can be invasive.

The beekeeping calendar

Your beekeeping year will largely depend on whereabouts you live and, of course, on the weather. A cold snap early in the year may find your bees starving and freezing into mid-spring, even if you live in a temperate area, and wet summers might make them languish as they try to avoid driving rain. In short, there is no such thing as a typical beekeeper's year, and as you gain experience you will be able to judge for yourself which actions will be necessary under various circumstances. The following are, therefore, guidelines only of the actions that you need to take, month by month, to keep your bees healthy and productive.

Spring

Spring is the time when the bees burst into life and get busy in the hive. It is an exciting period for the beekeeper as he must play cat and mouse with the weather, while hoping for a good crop of honey from his hive.

Early spring

The beekeeper's year begins in spring when the first flows of honey and pollen are matched by an increase in bee activity. The queen should be starting to lay eggs now, and the workers that overwintered should be beginning to forage. There will, however, still be plenty of cold days, and you should resist the temptation to open the hive, even if the bees seem to be flying happily.

Heft the hive to estimate how much food there is. Consider giving the bees a feed of syrup. If your feeder is on top of the crown board you can feed without letting any chill air into the hive, but avoid doing this on a dank, cold day. In a very cold spring you might need to think about feeding pollen or a pollen substitute.

Bees will fly on warm days, visiting apple blossom and storing pollen and honey. The numbers of bees will be gradually increasing because the queen will be laying. It is hoped that there are enough foraging bees and honey reserves to match the increasing burden of the new young.

- Check for signs of damage at the entrance to the hive and repair as necessary
- Change the floor of the hive, which will allow you to remove any rubbish and detritus and also to check for the presence of varroa mites
- If you find any signs of varroa infestation apply an appropriate treatment, for which you can get help from your local bee club or inspector

Above: *The first tentative inspection of spring. Choose a warm day and open the hive just for a short time.*

- Even though fruit trees should be coming into bloom check the weather regularly because the bees will stay inside in a cold snap and may need extra food
- Give a feed of syrup if necessary or of pollen or pollen substitute in very cold weather

Mid-spring

Wait for a warm day to have a proper inspection of the hive, but do not linger inside because the weather can change quickly. The purpose of this first inspection is to check for the queen and any signs of disease as well as to see how much food there is.

- Take the opportunity of a warm day to inspect the hive and remove any obvious rubbish, such as bits of comb
- Look at the honeycomb and replace any that has become dark with age
- Clean all your spare wooden frames so that they are ready to be reused
- Remove the mouse guard
- Check the queen excluder, making sure that the holes have not been filled with propolis

Late spring

The colony should be growing quickly by now and may well need extra feeding, so make sure that there is plenty of syrup available. If your hive is near to plenty of tree fruits or even an orchard be ready to add an extra super.

If you have not already done so, change the floor if you are not using an open mesh floor and flame the solid replacement before putting it in the hive to kill any over-wintering pests and diseases.

Because swarms can happen at any time from now on consider having an extra 'bait' hive available. This is also the time to think about the age and health of the queen and to decide whether you will re-queen the hive.

- Begin your regular weekly inspection of the hive
- Add a queen excluder and a super with fresh foundation
- Remove any mouse guards that are still in place
- Initiate your varroa treatment regime if you have not already done so
- Check the queen's health and take account of her age in deciding what to do about queen cells
- Be alert to the possibility of swarming

Above: *Clean old boxes with a flame from a blowtorch to remove the wax and kill disease.*

Checklist

- Look for the queen – is she healthy, laying well?
- Check the food stores – do you need to feed?
- Replace empty frames
- Check for varroa and treat as necessary
- Check for signs of swarming in late spring
- Only inspect the hive on warm days

The beekeeping calendar

Summer is the season with the most production in the apiary. The quantity of bees will have swelled, and they will be filling the hive with nectar they will never eat. Make sure you are ready with lots of frames to cater for the excess honey.

Early summer

This is the time to replace the central frames by moving them from the centre of the brood box to the outside of the box. In a laying cycle of 21–22 days these will have emptied and can then be completely replaced with new frames. You can reuse these frames once they have been thoroughly cleansed, preferably by freezing, or you can harvest the wax and make new foundation or candles or any of the other beeswax products.

A healthy queen should be at her peak of laying, and you should see busy workers tending eggs and grubs and plenty of capped brood. Make sure there is plenty of brood space available and consider adding another brood box. Make sure that you have some spare supers, and remember to be alert to the possibility of swarming. Depending on the age of your queen you might consider artificial swarming, queen replacement or simply removing queen cells.

- Place an extra super of foundation in the hive if you think the bees may need more room
- Check the brood box, which should be almost full of brood by now, to see if you can take some honey off
- Make sure you replace any frames or supers of honey you remove
- Check on the queen every week
- Be vigilant about swarming
- If your bees have been on or near oilseed rape remove the honey before it granulates on the comb
- Check regularly for varroa

Left: A common sight in summer, foraging bees vacuuming the nectar from every flower in the garden.

Midsummer

In most areas the threat of swarming should have passed by now, but you should remain vigilant. You can add an extra super for honey – which you will steal later.

- Check regularly for signs of varroa mite infestation
- Check the health of the queen, especially if she is two years old, when you might have to consider replacing her by supersedure
- Add extra supers as necessary but make sure that the bees are capping the cells in the lower super
- Prepare to extract the honey and make sure you have plenty of jars ready to keep it in

Late summer

This is the time of year when you can harvest the honey. Remove the supers so that you can begin to extract the honey, but make sure that you leave the colony with at least 15 kilograms (34 lb) of honey for their own supplies or you will have to feed them.

- Remove supers so that you can begin to harvest the honey
- Make sure that you feed your bees to compensate for the honey you have taken
- Remove wax and render it together with the wax you already have
- Clean and replace dirty frames
- Close the entrance to the hive to a strip about 5 cm (2 inches) or less wide so that the bees can withstand wasp intrusion
- If you can, move the hives to heather moorland

Checklist

- Check the queen is laying well – look for patch brood

- Check for varroa and treat as necessary

- Make sure there are plenty of frames to take lots of brood and honey

- If the weather is unseasonably cool for a prolonged period consider feeding

- Buy honey jars and prepare the extraction equipment

Left: *Make sure the extractor is serviced and cleaned, in a short while it will be needed.*

The beekeeping calendar

Autumn

Autumn is the time when you help the bees prepare for the winter. Hopefully, you should now have your honey, and so you need to make sure the bees are healthy enough to get through the winter.

Above: *Look out for wasps who may try to infiltrate the hive on their search for honey.*

Early autumn

As the temperature begins to fall and day lengths shorten you can start to limit your hive inspections to once a fortnight. You will have to start feeding any colonies that need it, and you should expect to notice a decrease in the number of bees in the brood. Feed the bees before the weather turns cold so that they have the opportunity to take it down and evaporate it into honey.

- Remove supers from the hive, clean and dry them thoroughly and store them safely for next year
- Combine colonies if necessary
- Test and treat for varroa when the honey has been removed
- Begin to offer a feed of sugar syrup
- Fit a mouse guard over the entrance
- Keep an eye out for wasps and robber bees

Mid-autumn

This is the time of year when wax moths can be troublesome in the hive. The larvae scoop out indentations in the wood of the frames or sides of the hive, where they spin a cocoon before pupating. They then move into the combs, and the moths lay eggs among the brood.

- Keep an eye out for infestations of wax moths
- Before closing the hive for winter give it one final, thorough inspection
- Make sure that all parts of the hive fit snugly together to keep out draughts so that the bees do not have to expend energy in filling gaps with propolis
- Fit a mouse guard if you have not already done so

Late autumn

By this stage of the year you will have stopped feeding the bees altogether. Now is the time to make sure that your hives are secure and can withstand the worst of any winter weather. If you live in a windy, exposed area you should consider adding a weight to the top of the hive to prevent it from being blown over.

- Try not to disturb your bees
- Add some insulation to the top of the hive in the form of a piece of carpet or a purpose-made 'quilt'
- Use a strap to make sure that the hive is securely held together to protect it from both vandals and bad winter weather

Checklist

- Are there enough food supplies – do you need to feed?
- Decrease the entrance to a small gap to deter mice
- Clean away old drones ejected from the hive
- Do not inspect as frequently and allow the bees to glue up the hive
- Make sure the hive is protected from bad weather

Left: *Anti varroa strips placed between the frames of a brood box.*

Winter

In winter the hive is asleep, the bees huddle together in their brood chamber and have glued themselves a warm space where draughts are kept to a minimum. But there is plenty for the beekeeper to do.

Above: *Give your equipment a good clean with soap and water as well as disinfectant.*

Early winter

Now is the time to catch up on all your records and to make sure that everything is brought up to date, especially if you have been applying varroa treatments. Take the opportunity to read and study, and inspect the outside of the hive only, but keep a watch for a build-up of dead bees and leaves at the entrance.

Winter is an excellent time to clean out and maintain your equipment. Make yourself a customized utility box and build your frames of foundation for next year.

- Check the hive regularly for storm damage
- Check the hive for signs of damage by woodpeckers
- If you have not already done so, strap and weigh down the colony so that it cannot be easily kicked over by vandals or animals or blown over by strong prevailing winds

Midwinter

Even though the weather may make you want to stay indoors, remember to check the outside of your hive regularly to make sure that it has not been damaged by falling branches or tiles. Bees will fly on warm days in winter, and if there is snow on the ground they may become confused by the extra ultraviolet light, so make sure that you brush away any snow from around the hive and erect a shade over the entrance.

- Brush away any snow from the entrance and from around the front of the hive
- Check hives regularly for storm and wind damage
- Make sure that the mouse guard is still in place
- Check that overhanging branches have not been damaged by storms and that they cannot drop on to the hive
- If you have a WBC-type hive look under the roof to make sure there is nothing hibernating there
- Look out for signs of flight on warm days, and observe any flying bees for vitality

Checklist

- Clean your equipment
- Build new frames and hives
- Read up on beekeeping
- Watch out for signs of attack by mammals – mice, badgers or people

Late winter

In late winter the queen will be beginning to lay again to produce the house bees that will take advantage of the first pollen flows of spring. This can a difficult time for your bees, but you must resist the temptation to open the hive. If you are really worried about starvation, get some help from an experienced beekeeper.

This is usually a quiet time of year, so take the opportunity to repair any spare hives, bring your records up to date and prepare frames with foundation.

- Check your hives for signs of storm damage
- Look out for signs of activity in the hive as winter-flowering bulbs come into bloom
- Make sure that you provide food for the hive
- Make sure that your bees have access to unfrozen water in a sunny spot

Right: *The beehive in winter is still: you would hardly know there were hundreds of bees alive inside the box.*

Recipes

If you are one of the many people who enjoy honey on toast for breakfast but then put away the jar when you are cooking, it's time to think again. Honey has been used to sweeten dishes from the times of the ancient Egyptians, and if you have your own supply you can use it in numerous dishes, ranging from simple breakfast cereals to sticky cakes and delicious ice cream.

Orange Flower and Honey Baklava

Prep time 40 mins, plus cooling
Cook time about 1 hour
Makes about 20 pieces

250 g (8 oz) blanched almonds
150 g (5 oz) blanched hazelnuts
75 g (3 oz) caster sugar
150 g (5 oz) unsalted butter, plus extra for greasing
50 ml (2 fl oz) extra virgin olive oil
22 sheets, about 300 g (10 oz), filo pastry (thawed if frozen)

SYRUP
300 g (10 oz) caster sugar
250 ml (8 fl oz) water
50 ml (2 fl oz) extra virgin olive oil
50 ml (2 fl oz) orange flower water
finely grated rind of 1 lemon
3 tablespoons clear honey
1 cinnamon stick

1 Make the syrup. Combine all the ingredients in a small saucepan and heat gently, stirring occasionally, until the sugar has dissolved. Increase the heat slightly and leave to bubble gently, without stirring, for about 20 minutes until it forms a light, sticky syrup. Remove the pan from the heat and allow the syrup to cool in the pan. Cover the pan and chill in the refrigerator, without straining.

2 Meanwhile, put the almonds, hazelnuts and sugar in a food processor or blender and process until the nuts are chopped but not ground.

3 Melt the butter with the olive oil in a heavy-based saucepan over a low heat and brush a sheet of filo pastry with the mixture, keeping the remaining pastry covered with a damp tea towel. Butter a 20 x 30 x 5 cm (8 x 12 x 2½ inch) tin and line it with the pastry, trimming it to fit the tin but leaving a little extra to allow for shrinkage during cooking. Repeat the process with 4 more sheets of pastry, brushing them with the butter and oil mixture as you layer them.

4 Scatter one-third of the nuts over the pastry and cover them with 5 more sheets of pastry, brushing each with the butter and oil mixture as you work. Add half the remaining nuts and cover them with 5 more sheets of pastry, the remaining nuts and then the remaining 7 sheets of pastry. Brush the top liberally with the remaining butter and oil mixture. Use a sharp knife to cut through the top few layers of pastry in a crisscross pattern to make about 20 diamond shapes.

5 Bake in a preheated oven, 190°C (375°F), Gas Mark 5, for about 40 minutes until golden. Remove the cinnamon and lemon rind from the syrup, and as soon as you take the baklava from the oven pour the syrup evenly over it. Leave to cool, then cut the baklava into the scored diamonds.

Honeyed Cereal

Prep time 10 mins
Cook time 20 mins
Serves 4

4 tablespoons sunflower or safflower oil
250 g (8 oz) clear honey
250 g (8 oz) millet, rye or barley flakes
250 g (8 oz) rolled oats
50 g (2 oz) sesame seeds
50 g (2 oz) dried peaches, banana slices, pears or
figs, chopped
125 g (4 oz) sultanas or seedless raisins
25 g (1 oz) coconut shavings
50 g (2 oz) pumpkin seeds
milk, to serve

1 Heat the oil in a small roasting tin, then stir in the
honey, flakes, oats and sesame seeds. Cook in a
preheated oven, 180°C (350°F), Gas Mark 4, for
20 minutes, stirring occasionally so the mixture
browns evenly.
2 Remove the tin from the oven and leave to cool. Mix
in the dried fruit, sultanas or raisins, coconut shavings
and pumpkin seeds. Store in an airtight container until
required, then serve with milk.

Honey Cinnamon Biscuits

Prep time 15 mins, plus chilling
Cook time 12–15 mins
Makes 20 biscuits

125 g (4 oz) plain flour
75g (3 oz) unsalted butter, diced, plus extra for greasing
¼ teaspoon ground cinnamon
3 tablespoons set honey

ICING
250 g (8 oz) icing sugar
1 tablespoon clear honey
lemon juice
pared rind of 1 lemon (optional)

1 Sift the flour into a bowl and rub in the butter. Add the cinnamon and mix in the honey to make a soft dough. Shape the dough into a roll about 20 cm (8 inches) long and wrap it in clingfilm. Chill until firm.
2 Lightly grease 2 baking sheets. Cut the dough into 20 slices and place them slightly apart on the baking sheets. Bake in a preheated oven, 190°C (375°F), Gas Mark 5, for 12–15 minutes. Leave on the baking sheets to cool slightly, then transfer to a wire rack to allow the biscuits to cool completely.
3 Make the icing. Sift the icing sugar into a bowl and gradually beat in the honey and enough lemon juice to make a thick glacé icing. Spread a little over each biscuit and leave to set. Decorate the biscuits with fine strips of boiled lemon rind, if liked.

Carrot, Honey and Sultana Squares

Prep time 10 mins, plus decorating
Cook time 20–25 mins
Makes 15 cakes

150 ml (¼ pint) sunflower oil, plus extra for greasing
3 eggs
125 g (4 oz) set or clear honey
50 g (2 oz) light muscovado sugar
200 g (7 oz) wholemeal self-raising flour
2 teaspoons baking powder
3 carrots, about 200 g (7 oz) in total, peeled and grated
100 g (3½ oz) sultanas

TO DECORATE
100 g (3½ oz) unsalted butter, at room temperature
finely grated rind and juice of ½ orange
200 g (7 oz) icing sugar, sifted
large and small candy-coated chocolate sweets

1 In a large bowl mix together the oil, eggs, honey and sugar. Add the flour and baking powder and mix well, then stir in the grated carrots and the sultanas.
2 Lightly grease and line a rectangular baking tin. Pour the cake mixture into the tin and smooth flat. Bake in a preheated oven, 180°C (350°F), Gas Mark 4, for 20–25 minutes until it is well risen and golden brown and the top springs back when lightly pressed. Leave the cake to cool in the tin for 10 minutes, then turn out on to a wire rack. Remove the lining paper.
3 Make the icing. Mix together the butter and orange rind in a medium-sized bowl. Gradually mix in the icing sugar and enough of the orange juice to make a soft, spreadable icing. Spread on the cake and cut into 15 squares. Decorate each square with the sweets.

Really Fruity Flapjacks

Prep time 15 mins
Cook time about 20 mins
Makes 15–20 flapjacks

100 g (3½ oz) unsalted butter
100 g (3½ oz) light muscovado sugar
5 tablespoons clear honey
375 g (12 oz) porridge oats
75 g (3 oz) ready-to-eat dried prunes, chopped
75 g (3 oz) ready-to-eat dried apricots, chopped
75 g (3 oz) raisins or sultanas
2 eggs, lightly beaten

1 Put the butter, sugar and honey in a small, heavy-based saucepan and heat gently, stirring. Remove from the heat and mix in the oats, prunes, apricots and raisins or sultanas until evenly mixed. Beat in the eggs.
2 Lightly grease a shallow 28 x 23 cm (11 x 9 inch) baking tin. Turn the mixture into the tin and level the surface. Bake in a preheated oven, 180°C (350°F), Gas Mark 4, for 20 minutes or until turning pale golden.
3 Leave in the tin until almost cold, then cut into fingers and transfer to a wire rack until they are completely cool. The flapjacks can be stored in an airtight container in a cool place for up to 5 days.

Pear, Cardamom and Sultana Cake

Prep time 20 mins
Cook time 1¼–1½ hours
Serves 12

125 g (4 oz) unsalted butter, softened
125 g (4 oz) soft light brown sugar
2 eggs, lightly beaten
250 g (8 oz) self-raising flour
1 teaspoon ground cardamom
4 tablespoons milk
500 g (1 lb) pears, peeled, cored and thinly sliced
125 g (4 oz) sultanas
1 tablespoon clear honey

1 In a large bowl mix together the butter and sugar until they are pale and light. Beat in the eggs a little at a time. Sift together the flour and ground cardamom and fold them into the creamed mixture together with the milk.
2 Reserve about one-third of the pear slices and roughly chop the rest. Fold the chopped pears into the creamed mixture with the sultanas.
3 Lightly oil and base-line a 1 kg (2 lb) loaf tin and spoon the mixture into the tin. Smooth the surface, making a small dip in the centre. Arrange the reserved pear slices down the centre of the cake, pressing them in gently. Bake in a preheated oven, 160°C (325°F), Gas Mark 3, for 1¼–1½ hours or until a skewer inserted into the centre comes out clean.
4 Remove the cake from the oven and drizzle over the honey. Leave to cool in the tin for 20 minutes and then transfer to a wire rack until it is completely cool.

Lavender Honey Ice Cream

Prep time 20 mins, plus chilling and freezing
Cook time 15 mins
Serves 4

6 tablespoons lavender honey
4 egg yolks
1 teaspoon cornflour
1 tablespoon caster sugar
300 ml (½ pint) milk
300 ml (½ pint) double or whipping cream
lavender sprigs, to decorate (optional)

1 Put the honey, egg yolks, cornflour and sugar in a bowl and whisk lightly to combine.
2 Put the milk in a heavy-based saucepan and bring it to the boil. Pour the milk over the egg yolk mixture, whisking well until it is combined. Return the mixture to the saucepan and cook gently, stirring, until the custard has thickened and coats the back of the spoon thinly.
3 Transfer the mixture to a bowl and cover with a round of greaseproof paper to prevent a skin from forming. Leave to cool, then transfer to the refrigerator to chill.
4 Transfer the custard to an ice cream maker and add the cream. Churn until frozen. Alternatively, beat the custard and cream together and put in the freezer, beating the ice cream after an hour to break down any ice crystals.
5 Serve the ice cream in glasses decorated with lavender sprigs, if liked.

Figs with Yogurt and Honey

Prep time 5 mins
Cook time 10 mins
Serves 4

8 ripe figs
4 tablespoons plain yogurt
2 tablespoons honey

1 Place the figs on a foil-lined grill pan and cook them under a preheated medium grill for 8 minutes, turning occasionally, until they are charred on the outside. Remove and cut them in half.
2 Arrange the halved figs on 4 plates and serve with a spoonful of plain yogurt and some honey spooned over the top.

Walnut and Honey Yogurt

Prep time 2 mins, plus chilling
Serves 1

1 dessertspoon honeycomb
100–200 g (3½–7 oz) low-fat, thick natural yogurt
6–7 walnut halves

1 Cut the honeycomb into pieces to release the liquid honey from the cells.
2 Stir the honey into the yogurt with the walnut halves and chill until ready to serve.

Hot Honey and Lemon Drink

Prep time 2 mins
Serves 1

2 tablespoons lemon juice
2 tablespoons set honey
boiling water

1 Mix together the lemon juice and honey in a mug.
2 Gradually pour in the boiling water, stirring to dissolve the honey, until the mug is almost full. Drink as hot as you can, but remember to allow it to cool for children.

Banana and Honey Shake

Prep time 2–3 mins
Serves 2

2 bananas
2 tablespoons clear honey
300 ml (½ pint) milk
2 ice cubes

1 Peel and slice the bananas and put them into a liquidizer or blender. Add the honey and milk and process until smooth.
2 Pour the mixture into individual glasses and add an ice cube in each. Serve with drinking straws.

Mead

Prep time 2 hours plus 1 year to mature
Makes approximately 5 litre bottle

5 litres (1.1 gallons) water
1.2 kg (2½ lb) honey
juice of 1 lemon
1 vitamin C tablet
1 teaspoon wine yeast

1 Put the water, honey, lemon juice and vitamin C tablet in a large, heavy-based saucepan and bring to the boil to kill the natural yeasts. (You can sterilize the mixture with the appropriate number of proprietary sterilizing tablets if you prefer.)
2 Leave the liquid to cool, then transfer it to a sterilized demijohn. Add the wine yeast and close with a sterilized airlock. The fermentation will take place for around 2 weeks and then the lees will start to settle.
3 Rack (siphon) off the liquid into another sterilized demijohn and store in a cool, dark place. (If you can manage to store it on a heavy stone floor the sediment will fall more easily.) Rack off the mead again.
4 When the liquid is clear transfer it to bottles and store for at least a year – if you can manage to resist it!

Glossary

Acaricide A preparation used to destroy mites.

Acarine A disease caused by mites (*Acarapis woodi*) that affects a bee's breathing tubes. It is sometimes known as Isle of Wight disease because it was first seen there in 1906.

Africanized bees Also, colloquially (if misleadingly), known as killer bees. A hybrid between the African honeybee (*Apis mellifera scutellata*) and several European honeybees. These bees, which have a reputation for being more aggressive in defence of their hives than European honeybees, have spread both north and south from Brazil, where the hybrids first occurred, and it is not known, given climate change, how much further they will spread.

American foul brood A notifiable disease, which may occur anywhere in the world, AFB is caused by *Paenibacillus larvae*, a spore-producing bacterium. The spores develop in the gut of larvae, which die in their cells, the cell caps appearing darker and sunken and often perforated. There is no known treatment, and infected hives, both stock and frames, must be destroyed. See also European foul brood.

Anther The part of a flower that releases pollen.

Bee brush A very soft brush or a large feather used to remove bees from the frame or to coax them into a confined space.

Bee cage A device for safely introducing a new queen into a hive. It is a mesh cage, which enables the other bees to lick the queen and pass her pheromones around the hive without harming her. The cage can be positioned between the frames of the brood chamber. The queen is kept inside the cage by a plug of newspaper or sugar candy, which the beekeeper can remove after a few days or allow the bees themselves to chew away.

Bee space The space of 7–8 mm (about ¼ inch) between the frames and walls of a hive that will not be filled with propolis (if smaller) or honeycomb (if larger), enabling the beekeeper easily to remove the frames. The space was first identified by the Rev. Lorenzo Lorraine Langstroth, a beekeeper in Philadelphia, in the early 1850s.

Beeswax A substance produced by eight glands on the underside of the abdomen of worker bees and used by bees to cap cells and build comb.

Bottom bee space hive A hive in which the frames hang in the boxes with a space between the bottom of the frames and the box. This style is most often seen in National-type hives. See also top bee space hive.

Brace comb A bridge of wax built between surfaces within the hive.

Brood The immature stage of the bee (eggs, larvae and pupae) in cells that may be uncapped or unsealed (eggs and larvae) or capped or sealed (larvae that will pupate into adult bees).

Brood box The part of the hive in which the queen is confined (by a queen excluder) and the brood is reared. A hive may contain more than one brood box.

Candy A solid food made from sugar, more frequently used to separate queens in a new hive.

Chalk brood A fungal infection, caused by *Ascosphaera apis*, which 'mummifies' larvae in the brood frame. Although it is occasionally confused with American foul brood, it is not as serious and may often be overcome by re-queening.

Clearer board A board used to remove bees from supers before the honey is harvested. The board will have one or more bee escapes, which allow bees to pass through but prevent them from returning.

Crown board Also known as the inner cover. The board that is placed on the top super and below the roof of the hive. The board has a hole in it for ventilation.

Drone A male bee. Unlike queen and worker bees, a drone has no sting.

European foul brood A notifiable disease, EFB is caused by the bacterium *Melissococcus pluton*, which infects the guts of developing larvae and competes with them for food. Not all infected larvae will die, and bees often eject the infected larvae themselves. However, if infected larvae survive to the pre-pupal stage, spores will be voided into the cells, infecting other larvae. A heavily infected hive should be destroyed, but in the early stages of infection shaking the swarm into a new hive is likely to be effective. See also American foul brood.

Exoskeleton The hard outside covering of an insect's body that protects and supports the insect.

Feeder One of two types are usually used, a large rectangular or smaller round version. They are filled with sugar syrup to feed the bees.

Flight board The wooden strip in front of the hive on which bees land.

Foundation A wax sheet embossed with a hexagonal pattern used as the basis for bees to build comb.

Frame A wooden or plastic structure that holds the wax comb and allows the beekeeper to remove the comb from the hive to inspect it. The beekeeper usually makes their own frames.

Guard bees Bees that wait at the entrance to the hive to protect it from foreign bees, wasps and animals. These bees release pheromones to alert the other bees in the hive if the colony is threatened.

Hefting The practice of lifting your hive with the roof on but with no supers in it as a way of establishing its weight. With practice, this enables the beekeeper to determine the weight of the colony and its food supplies.

Hive tool A metal tool used by beekeepers for levering and separating frames.

Hoffman frame A type of self-spacing frame.

Honey flow The period during which the supply of nectar from flowers is at its peak.

Landing board Also known as alighting board. The strip of wood that is attached to the bottom of the hive on which bees land before they enter the hive. It enables the beekeeper to see what bees are removing from the hive.

Langstroth hive The most commonly used type of hive in North America.

Mouse guard A grid allowing bees in and out of the hive while excluding mice.

Nasonov gland A gland on the abdomen of a worker bee that secretes a substance used to attract bees back to their hive.

National hive A square, single-walled hive; it is the most widely used hive in Britain.

Nectar The sugary substance that is produced by plants in order to attract pollinating insects and that is made into honey by bees.

Nosema A disease caused by a protozoan (*Nosema apis*), which affects the gut of adult bees.

Nurse bee The name given to a worker bee that helps to rear brood in the hive.

Oilseed rape The yellow flowers of this important agricultural crop, *Brassica napus*, are a good source of pollen, but the honey produced crystallizes quickly and sets so hard that bees cannot get it out of the combs. If your bees are near crops of OSR you must be ready to take the honey from the hive early in the season. Also bear in mind that farmers are likely to apply pesticides to their OSR crops, which will not only kill wild bees but could well destroy your own colony.

Open mesh floor A fine mesh floor that is used to both improve ventilation in the hive and help control varroa mites, which fall through the mesh but are unable to return to the hive.

Pistil The female reproductive organ of a flower.

Pollen The dust-like grains that are produced by a flower's anthers (the male part of a flower) and that are used to fertilize the female ovule.

Pollen basket Also known as pollen sac. The area on the hind leg of a bee in which pollen is transported to the hive. The basket, actually a hollow in the tibia, is surrounded by brush-like hairs, which enable the bee to scrape the pollen caught on its body hairs into the basket.

Porter bee escape A one-way bee escape, often used on clearer boards, that allow bees to exit but not return.

Propolis Also known as bee gum or bee glue. The reddish or black-brown resinous substance that is collected by bees from plants, such as from the buds of *Aesculus hippocastanum* (horse chestnut), and used in the construction of the hive.

Queen The sexually developed female bee that lays eggs.

Queen cell A large cell on the face or bottom of the frame that houses a grub destined to become a queen.

Queen excluder A screen with slots or a mesh that allows worker bees to pass through it but can be positioned to exclude the queen and drones from parts of the hive.

Queen marking grid A device for holding a queen on the frame surface allowing you to mark her. Marking sets the queen's age and makes her more visible.

Robbing The taking of honey from a hive by wasps or bees from another colony.

Royal jelly Also known as bee milk. The substance secreted by the worker bees and fed to future queens.

Sac brood A viral disease that causes larvae to die before their final moult. Re-queening is usually effective.

Skep An old-fashioned type of hive, made of wicker or straw and without movable frames, which the beekeeper can use to take swarms.

Smoker The device used by beekeepers to introduce cool smoke into a hive to calm bees before the hive is opened. Suitable materials to burn include wood shavings, newspaper, egg boxes, old hessian sacking, dried conifer clippings and dried grass cuttings.

Stigma The part of a flower's pistil that receives the pollen during pollination.

Sting Queen and worker bees, but not the drones, have a barbed sting. When the sting is used, it is left in the wound together with parts of the bee's viscera. This means that after the bee has stung it will die.

Sugar syrup A solution of 50:50 sugar : water, used to feed bees in times when no nectar is flowing or honey supplies are low.

Super The chamber or chambers above the brood box in which honey is kept.

Supersedure The process of removing a queen and replacing her with a new one without the colony swarming.

Sulphur candle A device for killing off a whole colony if some disease problem makes it necessary.

Swarm A mass of bees that is not in a hive. The bees could be attempting to establish a new colony or escaping from an unsatisfactory hive. Ideally, a swarm includes a mated queen.

Top bee space hive A hive in which the frames are suspended so that there is a space between the top of the frames and the top of the box. This arrangement is most often seen in Langstroth-type hives. See also bottom bee space hive.

Varroa destructor The mite that breeds in sealed brood cells.

Varroa strips A plastic strip impregnated with insecticide that will selectively kill varroa mites in the hive.

Varroosis The disease of bees caused by the parasitic mite *Varroa destructor*.

Waggle dance A circular or figure-of-eight dance performed by the bees. They use this movement to communicate with other bees the distance and direction of a food source from the hive.

Wax moths Both the lesser (*Achroia grisella*) and greater (*Galleria mellonella*) wax moth are serious pests in the hive, where they damage stored comb. The larvae manage to scoop out indentations in the wooden sides of frames and hives (including those that are in storage) where they pupate.

WBC hive A double-walled hive, designed by William Broughton Carr in 1890, which is often included in gardens because of its attractive, pagoda-like appearance.

Worker An immature female bee. Worker bees are the most numerous residents of a hive.

Glossary

Index

Acknowledgements

Special thanks to Rachel Graham, Maggie Bohme, John Gresty, Keith Hibbert, Derek Hartington, Ian Molyneux, Manchester and District Beekeepers Association (www.mdbka.com)

Executive Editor Jessica Cowie
Editor Kerenza Swift
Executive Art Editor Leigh Jones
Designer Ginny Zeal
Senior Production Controller Simone Nauerth
Picture Researcher Sarah Smithies

Picture Acknowledgements

Commissioned Photography:
© Octopus Publishing Group Ltd/Paul Peacock

Other Photography:
Alamy/Renee Morris 6; /Ros Drinkwater 5 top centre, 41.
Clive Nichols 42, 43 right.
Corbis UK Ltd/Engquist Etsa 43 left.
Fotolia/Kharr 32; /Mikhail Tolstoy 97;
/Philippe Simier 61.
Getty Images/Michael Durham 11.
istockphoto.com/Karen Massier 44; /Torsten Karock 62.
Leigh Jones 32, 35, 54, 63, 125, 135.
Octopus Publishing Group Limited/Gareth Sambidge 133;
/Gus Filgate 127; /Ian Wallace 131.
Rachel Green 15.
Shutterstock/Adam Tinney 1; /Hway Kiong Lim 19;
/Mike Tolstoy 102.
The Garden Collection/Roger Benjamin 60.
The Trustees of the British Museum 13.
Zachary Huang 20, 28, 93; /M V Smith 95.